eco
LOGICAL!

JOIN THE DEBATE! – ALL THE FACTS AND FIGURES, PROS AND CONS YOU NEED TO MAKE UP YOUR MIND

JOANNA YARROW

WITH CALEB KLACES

DUNCAN BAIRD PUBLISHERS

LONDON

eCOLOGICAL!

JOANNA YARROW WITH CALEB KLACES

First published in the United Kingdom and Ireland in 2009 by
Duncan Baird Publishers Ltd
Sixth Floor
Castle House
75–76 Wells Street
London W1T 3QH

Conceived, created and designed by
Duncan Baird Publishers

Editor: James Hodgson
Designer: Luana Gobbo
Managing Editor: Christopher Westhorp
Managing Designer: Daniel Sturges
Commissioned artwork: Peter Grundy

British Library Cataloguing-in-Publication Data:
A CIP record for this book is available from the British Library

ISBN: 978-1-84483-822-6

10 9 8 7 6 5 4 3 2 1

Typeset in Meta
Colour reproduction by Colourscan, Singapore
Printed in Malaysia by Imago

Note:
This book was printed using vegetable-based inks on paper from
paper mills that meet environmental standard ISO 14001

CONTENTS

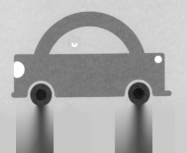

INTRODUCTION

Ecological: of or relating to ecology; beneficial to or not harmful to the environment (*Penguin English Dictionary*).

Most of us would prefer to live in a way that protects rather than damages the planet. But since almost every aspect of our lives has some kind of environmental impact, it can be hard to know where to start. What's the most logical way to lead happy, fulfilling lives in balance with the environment?

The rise of ecology

Not very long ago, ecology was seen as a special-interest subject of little relevance to most people's day-to-day life. Media coverage occasionally touched on distant-sounding ecological horror stories – rainforests being destroyed or exotic species disappearing. Or they beguiled us with cute images of cuddly creatures

– the charismatic megafauna that had us reaching for our wallets to donate to conservation charities. Some of us enjoyed communing with nature on country walks or feeding the birds in our backyard. But ecology still seemed more a hobby than a life-or-death issue. Something to be enjoyed at the weekend and then forgotten about amid the hurly burly of "real" life.

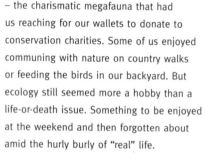

But over the last few years the perilous state of our world has become ever more apparent. Year by year, the natural systems that our very existence depends upon have shown themselves to be in increasingly dire straits.

Atmospheric CO_2 concentrations are growing at the fastest rate ever recorded[1], and the impacts of climate change (widely

accepted to be caused by rising CO_2 levels from human activity) are manifesting themselves years earlier than anticipated: scientists are predicting ice-free summers in the Arctic between 2011 and 2015 – 80 years earlier than predicted as recently as 2007. When this happens ships will be able to sail between the Atlantic and the Pacific without going through the Panama Canal or around Cape Horn for the first time in recorded human history. An increased incidence of drought threatens nearly 2 billion people who rely on livestock grazing for their livelihoods. Globally, nearly a quarter of all mammal species and a third of amphibians are under threat, largely – it is believed – as a result of human behaviour. Scientists fear that the planet could be about to suffer the biggest round of extinctions since the death of the dinosaurs[2].

As the systems we rely upon reveal their limits, it's become less easy to dismiss the natural world as a mere backdrop for our lives, to be enjoyed or ignored at whim. Suddenly we have to stop thinking of ourselves as set apart from nature, and realize that we are part of, and entirely dependent upon, the natural world.

Years of rapid technological advances have given us comfort beyond our ancestors' wildest dreams. But much of this development has happened without consideration for the long-term impact on the resources that fuel it or the ecosystems that must absorb its by-products and waste. We've neglected the joined-up thinking that asks whether our innovations are sustainable. We've not been logical.

No one is suggesting that we must revert to cave-dwelling hunter-gatherers, but if we're to design healthy, fulfilling lives for a growing population with high expectations, we need to start recognizing that we're part of the world's ecology – the interrelationship of living organisms and their environments. We need to start living in ways that respect rather than ignore our relationship with the natural world – in ways that are truly "eco-logical".

[1] WorldWatch Institute
[2] Global Environmental Outlook 4 (2007)

The challenge: a new way of thinking

As the scale of the challenge becomes increasingly apparent, environmental problems are rarely out of the headlines. And commentators, pundits, campaigners and nay-sayers are bombarding us with their opinions on the nature of the situation and suggestions for how we might best avert environmental Armageddon.

For most people, this will require a new way of thinking. Our lifestyles have been designed in ways that distance us from the impact of our actions – our rubbish is taken far away rather than piling up in our backyard, and we rarely see the circumstances in which our food is produced or our electricity generated.

The huge volumes of information and misinformation can make it hard to see the wood from the (endangered) trees when it comes to understanding exactly what's going on and to deciding how we should react to these daunting times.

About this book

This book isn't intended to be a textbook on sustainability. Instead it's designed as a guide to help you navigate some of the key environmental issues of our time, to address some of the common misconceptions that still crop up in environmental discussions, and to spell out the main dilemmas we'll have to resolve if we're really to live more ecologically.

The book is divided into five chapters:

Forces of change sets the scene, examining some of the fundamental issues that shape our lives – from climate change and oil supply to population growth and waste management.

The way we live looks at where and how we conduct our everyday lives and asks how we can continue to live comfortably while protecting the interests of the natural world. Topics range from low-carbon homes and the pros and cons of city living to the ethics of profitable business and the ecological footprint of leisure.

Power to the people assesses lower-carbon alternatives to traditional fossil-fuel power generation, weighing up options such as Carbon Capture and Storage, nuclear power and the various forms of renewable energy.

World in motion considers how we can satisfy our wanderlust without destroying the planet we set out to explore, comparing last century's carbon-fuelled modes of transport with cleaner, smarter 21st-century alternatives.

Eat, drink, shop examines how we can feed and clothe ourselves sustainably, how supermarkets stack up against local, independent retailers and how we can put our consumer power to good purpose.

Each chapter contains a series of eco-dilemmas. These are examined from different perspectives, with case studies and examples to illustrate the nature of the debate and point to possible solutions.

I hope that this book will help stimulate you to find ways of living that make sense for you and the planet. Climate change and the other ecological challenges we face won't go away, but we have a window of opportunity in which to address them. Let's try to do that in the most logical way!

Joanna Yarrow

CHAPTER 1

FORCES OF CHANGE

Twenty-first-century life is shaped by a complex set of often conflicting pressures – expanding populations and dwindling resources, growing industrialization and shrinking rainforests, a changing climate and a developing world. We are bombarded with predictions of impending doom, countered by assurances that there's nothing to worry about. Who should we believe? What can we do to make a difference?

This chapter tackles the key issues:

- How do we know it's getting hotter and what are we going to do about it?
- Is the global economy fit for purpose?
- When will we run out of oil?
- Is there room for all of us?
- Do China and India's growing carbon emissions make our eco-efforts futile?
- Will we drown in our own waste?
- How can we relieve water stress?
- And do we really need to worry when species disappear?

CLIMATE CHANGE

A stable atmosphere is crucial for supporting life on Earth. But in recent years that balance has been put out of kilter: by burning fossil fuels in ever-increasing amounts we're releasing **unprecedented levels of** CO_2 into the atmosphere. **The Earth is getting hotter,** the climate is changing and **debate rages** about how we should address what is now probably the biggest challenge of our age.

⬆⬇ We need to take radical action immediately

- The change in climate is unprecedented and has been induced by humankind's carbon-heavy lifestyle – only a **massive and immediate reversal** in this can halt the heating up of the planet
- The Intergovernmental Panel on Climate Change (IPCC) estimates that by 2100 the planet will heat up 1.1–6.4°C (2.0–11.5°F) – we have only a short window of opportunity to take action before the problem becomes **impossible** to solve
- Catastrophe looms: as well as causing dangerous rises in sea levels, extreme weather patterns, floods and drought, a 2°C (3.6°F) increase in global temperature is thought to be a **"tipping point"** beyond which we may be unable to influence climate change

THE DIFFERENCE BETWEEN CLIMATE AND WEATHER

It's important not to confuse climate with weather. Weather varies over the short term, and can be changeable and unpredictable from hour to hour, day to day, week to week or season to season. Climate is the average of weather over time and space. Natural changes in climate are typically very gradual, taking generations, centuries or even millennia for even small variations to become the norm. However, over the last few decades, climates have been changing faster than flora, fauna, people and places can adapt.

! There's currently a greater concentration of greenhouse gases in the atmosphere than there has been for up to 800,000 years, and they are increasing by around 2.5% each year.

Some change is inevitable, so we must adapt

- Even if emissions are reduced dramatically, CO_2 already released through human activity will have climate impacts for many decades to come, meaning that some **further climate change is inevitable**
- Even under the most **optimistic scenario**, it will take time to phase out fossil fuels, so carbon **emissions will continue to rise**
- It looks increasingly unlikely that warming will be limited to 2°C – so as well as reducing emissions, we need to **focus on damage limitation** and adapt to a changing climate

AT WHAT PRICE?

The economic impacts of climate change are hard to quantify, but the consensus is that acting immediately will cost far less than waiting. Sir Nicholas Stern argues that doing nothing now to address the issue could shrink global GDP by around 20%. In contrast, taking action to avoid the worst effects of climate change immediately could cost as little as 1% of global GDP.

Source: *Stern Review of the Economics of Climate Change* (2006)

FOCUS: Temperature on the rise

Since the Industrial Revolution CO_2 levels have increased by 35%. Over the same period global average temperatures have risen by almost 1°C (1.8°F). However, the precise nature of the link between CO_2 and temperature is complex and unclear.

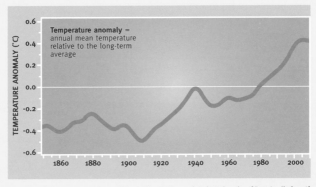

Temperature anomaly = annual mean temperature relative to the long-term average

Source: Climatic Research Unit, University of East Anglia (2008)

Is the planet warming?

"Greenhouse gases" (GHGs) in the Earth's atmosphere absorb and reflect infrared radiation. The more they build up in the atmosphere, the less heat escapes out into space and the warmer the Earth becomes. For most of history this has been a good thing, making the planet warm enough to sustain life.

Up to half of the GHGs produced by the planet's ecosystems have been reabsorbed in natural carbon "sinks" such as plants and the oceans. Today, however, the concentration of CO_2 is the highest it's been for 800,000 years and these sinks simply can't keep up.

As well as the direct impact that rising temperatures have on ecosystems, extra heat means extra energy in the climate system, which can lead to more volatile, more violent and less predictable weather of all kinds, with more storms, droughts and floods.

How do we know how things are changing?
Although some sceptics dispute the data showing a warming trend, citing flawed modelling and margins of error, the vast majority of climate scientists refute these critiques and point to clear signs in the world around us. Data about weather patterns in the distant past has been

> ⚠ **Sea levels may rise by 2m (6.6ft) this century – just a 1m (3.3ft) rise would put dozens of big cities at risk.**
> Sources: IPCC/Potsdam Institute/Sir Nicholas Stern

CLIMATE IN CONTEXT

It is true that the planet has been this hot before, even hotter in fact. There have also been similar levels of carbon in the atmosphere before. However, never before have GHG concentrations changed so quickly; in less than 20 years the biosphere has altered in ways that have previously taken 1,000 years or more. Most life-forms on the planet have evolved to be tuned to the temperature range that has prevailed until now, and the rate of change we're currently experiencing is leaving them little time to adapt.

What about the sun?
Is climate change caused by solar activity? Changes in the sun's intensity alter the Earth's temperature only very slightly. It is believed sunspots (or their absence) contributed to around a 0.3°C (0.5°F) fluctuation in global temperature in the past 1,000 years – much less than the rise in average temperature over the last century.

obtained by studying "ice cores" – cylindrical sections sampled from an accumulation of layers of snow and ice. The different hydrogen isotopes the ice cores contain indicate the Earth's changing temperatures over time (heavy isotopes were retained in warmer periods), and tiny bubbles reveal ancient concentrations of GHGs. Ice cores show that as the concentrations of GHGs rise so do temperatures, and when GHGs fall, cooling is amplified. This complex feedback system linking concentrations of GHGs to temperature has driven long cycles of warming and cooling over millennia. But the rate at which CO_2 concentrations are now rising – and the speed at which temperatures are consequently predicted to rise – is unprecedented.

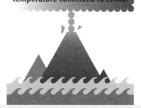

KOOL KRAKATOA

Sulphur in the ash released from volcanic eruptions reflects sunlight back into space, but like solar activity the effect is tiny compared to global warming from human activity. The cooling effect caused by the Indonesian volcano Krakatoa's explosion in 1883 lasted a few years, and then the planet's temperature continued to climb.

▶ FOCUS: A gassy problem

Human activity releases around 700 tonnes of GHGs per second into the atmosphere. This is faster than they can be naturally reabsorbed, so they build up, trapping heat and warming the Earth. Several GHGs contribute to climate change, notably CO_2, which is emitted in the largest quantities and remains in the atmosphere for the longest. Water vapour, a gas not generally emitted directly by human activity, also contributes to the warming effect.

Greenhouse gas	Contribution to man-made climate change relative to CO_2 over its lifetime	Estimated relative overall contribution to man-made climate change (%)	Atmospheric lifetime (years)
Carbon dioxide (CO_2)	1	60	50–200
Methane (CH_4)	c.25	15	c.10
Nitrous oxide (N_2O)	c.300	5	c.150
Tropospheric ozone (O_3)	c.30	8	Weeks to months
Hydrochlorofluorocarbons (HCFCs)	Up to 15,000	12	60–100
Water vapour (H_2O)	–	Unknown	Days

What should we do about it?

How much do we need to change?

Since the Industrial Revolution, atmospheric CO_2 levels have increased by 35% and global average temperatures have risen by almost 1°C. Scientists agree that atmospheric CO_2 levels need to be kept below 350 parts per million (ppm) to stabilize the climate. But atmospheric CO_2 levels have stayed higher than 350ppm since early 1988 and continue to rise each year. We need to reverse this trend, and fast. There is international consensus that to keep any further average temperature rise below a "safe" 2°C (see Six Degrees of Warming, right) world CO_2 emissions must be stabilized by 2015 and there must be a rapid trajectory down to almost zero by 2050.

There is no magic solution or single way of addressing climate change. It will require concerted action by international bodies, governments and businesses – as well as individuals – to minimize carbon emissions from every aspect of human activity.

The race is on to shift to a decarbonized economy within which low-carbon ways of living can deliver a decent quality of life. We must start using energy much more efficiently, and fast-track renewable energy production worldwide. Fossil-fuel consumption must be dramatically reduced and any that is used must be burned in the cleanest and most efficient way possible, and unavoidable emissions captured and stored safely. Deforestation (currently contributing some 17% of

SIX DEGREES OF WARMING

The global impacts predicted with each degree of warming:

1°C The Great Barrier Reef is nearly dead. Drought devastates US farmland. Monsoon in North Africa makes the Sahara lush.

2°C A third of all species alive today become extinct. Temperatures in Europe match the Middle East.

3°C A "tipping point" is passed and climate change spins out of control as a series of "positive feedback" reactions are triggered. Amazon rainforest burns down. Millions of climate refugees move north.

4°C No ice at the North Pole for the first time in 3 million years. Seas rise by up to 5m (16ft). Deserts in southern Europe.

5°C The South Pole has no ice. Forests grow in Antarctica. Frequent tsunamis and hurricanes.

6°C The last time Earth was this hot, 95% of all species previously alive were dead.

Source: *Six Degrees: Our Future on a Hotter Planet*, Mark Lynas (2007)

annual greenhouse-gas emissions) must be halted to protect one of the planet's greatest natural defences against climate change.

Elemental engineering

Even if we were to completely halt the production of man-made carbon emissions, we'd still experience some climate change as a consequence of the excess greenhouse gases that have already accumulated in the atmosphere, so in the process of reducing emissions we'll need to adapt to some unavoidable warming. One major impact will be rising, warmer seas with more extreme storm surges, bringing erosion, flooding and extreme weather. Half the human population lives within 200km (125 miles) of the coast, so governments will need to improve flood and coastal defences, and barrier, levee and drainage systems. More extreme weather conditions will also require fundamental re-engineering of everything from city design and transport infrastructure to agriculture worldwide.

> ❗ **One way to ease the warming problem could be to sequester carbon by enhancing the planet's natural carbon "sinks" (an artificial equivalent is Carbon Capture and Storage, see page 69). Preserving existing forests and peat bogs as well as massive planting of trees will play a key role. More radical is a suggestion to encourage large amounts of plankton or algae to bloom on the ocean surface. When that sinks to the ocean floor it could lock away carbon. However, interfering in natural systems in this way can be counterproductive.**

NATURE AND ADAPTATION

As the world has warmed many animal species have responded to the resultant changes by adapting their migratory behaviour. Temperature rises and habitat changes mean that up to 500 million birds travelling between African winter quarters and breeding grounds in Europe and Asia are having to fly significantly further each year. Requiring more energy to travel, the added distance is a significant threat to their survival. Similarly, many types of fish seem to have moved further north, according to evidence from trawlermen.

ECO-NOMICS

The ecological cost of the things we buy isn't usually included in the price tag. As the ecosystems that every aspect of our economy depends upon are pushed to breaking point, should we develop a new kind of "eco-nomy" that incorporates ecological accounting?

⬇⬆ If it ain't broke ...

- "Green" economics is a very different system from the one we currently use to value (or rather, price) things – the shift would be monumental, with plenty of opponents who have a vested interest in the current system

- In an increasingly global economy, how will we decide who pays for environmental costs?
- Some economists believe that ecological problems can be addressed within the current system by means of taxation

⬇⬆ Nature must have its price

- The current system's short-term outlook only exacerbates environmental problems
- There are now fair and practical ways of assigning responsibility for environmental impacts
- Excessive subsidies of unsustainable practices distort the market and discourage any move toward more sustainable ways of doing things
- A green economy could generate significant employment in truly sustainable jobs

THE NATURE CRUNCH

In autumn 2008 the global financial sector lost between US$1 trillion and US$1.5 trillion due to the "credit crunch". In October 2008 a European study on ecosystems reported that we're losing natural capital worth between $2 trillion and $5 trillion every year as a result of deforestation alone.

Source: *The Economics of Ecosystems & Biodiversity* (2008)

Natural capital and externalities

We're used to paying a price for resources or "products" taken from the natural world, such as timber, oil, coal, minerals or fish. But traditional economic systems don't put any value on ecosystems, such as wetlands, prairies, coral reefs and rainforests, or ecosystem services, such as the ability of nature to assimilate waste, turn sunlight and water into edible plants, and create oxygen, upon which all aspects of our existence depend.

These systems and services are seen by economists as "externalities", which means they aren't included in the price we pay – and because they're given no value there's little economic incentive to conserve or protect them. But something or someone must pay the cost of, for example, cleaning up the soil degradation and water pollution caused by heavy fertilizer use in the production of a beefburger – even if that cost isn't included in the price on the menu.

So how much is this natural capital worth?

It's very hard to work out the value of ecosystem services, as we don't fully understand – or even know about – many of the services that nature provides. One study has estimated natural services to be worth US$33 trillion per year (meanwhile "gross global product" is around US$18 trillion per year). But since services such as the production of oxygen can't be replaced by technology, in practice they're priceless.

KEEP IT LOCAL

Green economies start small and grow

Communities around the world, from Totnes in southwest England to Berkshire County in Massachusetts, are developing micro-economic systems that keep wealth in the community and reduce food and trade miles. Some issue special currencies that can be spent only in local businesses; others do away with cash completely.

LETS (Local Exchange Trading Systems) are a modern-day form of barter in which people exchange goods or services directly. People earn LETS credits by providing a service, and can then spend the credits on whatever is offered by other members.

These initiatives aren't perfect – many fold because of the strain of administering them – but they offer a potential model for a more sustainable, community-centric way of organizing economies.

What's money for?

Whatever makes you happy

Every year the Happy Planet Index ranks countries according to life-satisfaction surveys and sustainability indicators. It consistently finds that, beyond the point at which a reasonable standard of living can be paid for, increasing income and consumption do not lead to greater well-being. In fact, as people earn and spend more, greater materialism, a diminished sense of community, and destruction of natural capital often combine to reduce well-being. And good lives don't have to cost the Earth: for example, Germans and Americans report similar levels of life satisfaction, yet the average German's ecological footprint is half the size of the average American's.

Wise investments

A growing number of banks and investment funds offer accounts that restrict their investments to organizations that meet a set of environmental or ethical criteria.

Ethical funds fluctuate like any other investment, but some financial analysts believe that they may represent a sounder long-term investment than many conventional funds. This is because qualifying companies tend to have sustainable business models founded on the reduction of waste and the cultivation of lasting relationships with employees and shareholders, all of which helps them to perform better over the long haul than their competitors.

A QUESTION OF PRIORITIES?

Far more is spent on defence than on protecting ecosystems.

Some world military budgets are as follows:

US	$560 billion
UK	$59 billion
France	$53 billion
China	$50 billion

Source: Stockholm International Peace Research

In contrast, the Earth Policy Institute's "Plan B" budget estimates the additional expenditures that would be needed to meet key social and environmental goals, including the following:

• Universal primary education $12 billion
• Eradication of adult illiteracy $4 billion
• Universal basic health care $33 billion
• Reforesting the Earth $6 billion
• Protecting topsoil on cropland $24 billion
• Restoring fisheries $13 billion

A Green New Deal

The year 2008 saw a "triple crunch" – a global banking crisis, increasing concerns about the climate and erratic energy prices. Could a "Green New Deal" drag the world out of this financial trough and into a greener, richer economy?

Drawing inspiration from Franklin Roosevelt's New Deal, which tackled the 1930s depression and helped set up the world economy for the unprecedented growth of the second half of the 20th century, a growing number of world leaders believe that reformulating transport, energy generation, industry and farming along ecological lines will generate massive "green collar" employment and equip the world to face the unique challenges of the 21st century.

The proposed Green New Deals reconcile growth with green by using what resources we have more efficiently and realizing value from maintaining ecosystem services, rather than depleting them.

PRICING NATURE

In the current economic framework, it's unreasonable to expect countries that host important ecosystems such as rainforests not to exploit the natural capital within their borders – particularly as these countries are often among the world's poorest.

In an enterprising move which reconciles economic and environmental considerations, Ecuador has asked rich countries to pay it US$350 million a year as compensation for not extracting 1 billion barrels of oil in the Amazon rainforest. And in UN talks on a new climate treaty, more than 190 nations are considering a plan to pay tropical nations billions of dollars a year to leave forests alone.

WHAT DRIVES US?

Taxation offers another route to positive change. Taxing an ecologically undesirable activity, such as driving, can help dissuade people from doing it, and as long as attractive alternatives are offered (such as efficient public transport) this can help achieve a shift toward more sustainable activity.

A 2008 US survey illustrates the important role that cost plays in influencing behaviour. The study found that more than 24 million Americans – 11% of the adult population – are using public transportation more than they did in 2007. Nearly one in three respondents (32%) said their biggest motivator to choose public transportation over driving would be high fuel prices. A long way down the list was concern for the environment (4%).

Source: HNTB Companies (2008)

PEAK EVERYTHING

It is widely believed that the world will soon reach "peak oil", after which production will **decline** and prices **increase**. Evidence also suggests that we're using potentially renewable resources such as soil, freshwater and fish **faster than they're replenished**. Historically, a few societies have managed to prosper despite scarce resources, but many disappeared because their stocks ran dry. **Which way will we go?**

⬇⬆ There's nothing to be gained from waiting

- History has shown that societies can **collapse** when they run out of sources of energy and materials
- Intensive farming is **turning fertile land into desert** in many parts of the world and other resources such as **freshwater** are under increasing pressure
- Predictions for Peak Oil are rapidly becoming more specific – and it seems very likely that **the turning point is close** if we haven't passed it already
- The influential Hirsch report for the US Department of Energy suggested it will take 20 years to make a transition **beyond an oil economy**

PEAK SOIL

It takes tens of thousands of years to make topsoil 15cm (6in) deep. Worldwide, human activity is depleting soil 10 to 20 times faster than it's being replenished. About 10% of the energy input for agriculture is now used in offsetting the negative effects of soil erosion. As most of this energy is produced using fossil fuels, dealing with declining soil is also diminishing oil.

❗ More than half the world's original forested area has been converted to other uses. At present rates, a quarter of what's left will be converted in the next 50 years.

⬇⬆ Not time to panic yet?

- Predicting resource depletion is a complex business – estimates of what's left vary, so **why should we** base all our plans on the most pessimistic projections?
- Some increases in the price of resources have more to do with **short-term** political and economic disturbances than with a **terminal** decline in supply
- As a resource becomes scarcer and its **price rises**, people tend to change their behaviour so that they **use less** of the resource

CRITICISMS OF PEAK OIL THEORY

Opponents of Peak Oil theory (see page 24), believe it gives too pessimistic a view. They argue that Peak Oil calculations don't take into account the potential for discovery of new oil fields or the likelihood that future technological advances will make it possible to extract oil at a faster rate than the current estimate. Critics also point out that there have been times in countries such as Iraq, Nigeria and Venezuela when political, rather than geological, problems have impeded oil production – thereby skewing Peak Oil estimates.

▶ FOCUS: How much is left?

We might not completely exhaust all possible fossil fuel reserves for another few hundred years at current usage rates, but our energy consumption is set to double by 2050, and meanwhile supplies of oil are close to peaking and the remaining reserves are becoming increasingly difficult and expensive to extract.

In 2008 the chief economist at the International Energy Agency predicted that crude oil production could peak by 2020. Some in the industry believe we'll hit that point by 2013 at the latest, and others fear it may already have been reached. Increasing scarcity further, oil companies across the world have cut investment in new projects in the wake of the economic downturn since 2008.

❗ The US Energy Information Administration issued no warning before the domestic supply of natural gas failed to meet demand after 2000. Prices quadrupled.

HOW MANY YEARS HAVE WE GOT LEFT?

 OIL: 20–40 years

 GAS: 50–170 years

 COAL: 200–400 years

Sources: Oil & Gas Journal/Energy Information Administration/World Energy Council

A short history of collapse

The downfall of the civilization on Easter Island in the late 17th century has been cited as a warning of what can happen when a society overexploits the resources it depends on. The story goes that the community chopped down all their trees, which led to soil erosion, starvation, cannibalism and then complete collapse. However, a rival explanation puts the blame at the door of the Europeans who first visited the island in the 17th century and who not only murdered much of the population but also unwittingly introduced diseases to which natives had no resistance.

Whatever the truth in this case, there are plenty of other historical examples of societal collapse (such as the Maya in central America and the Anasazi in what is today New Mexico) for which environmental change is the most compelling explanation.

FOGGY FORECASTING

Predicting oil prices is not an exact science. In 2004 OPEC published its crude oil price forecast for the next two decades – a steady decline to US$20 per barrel by the year 2025 – just a few months before the price took off on a climb past US$100 a barrel for the first time.
Source: OPEC (2004)

▶ FOCUS: How does Peak Oil theory work?

Peak Oil projections are based on the rate at which new oil reserves are discovered within a given oil field and the rate at which the oil is extracted. Production tends to follow a bell curve, known as a Hubbert curve after the research geologist M. King Hubbert, who in 1956 predicted that US oil production would peak between 1965 and 1971 (see right). Sure enough, it crested in 1970.

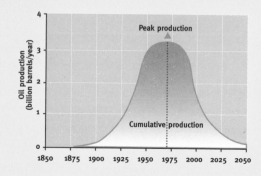

Can we survive peak oil?

Does the declining availability of vital resources spell inevitable apocalypse? It would if we tried to carry on with business as usual, so adaptation is the key to survival.

After the fall of the Soviet Union, Cuba lost more than 50% of its oil imports and 85% of its trade economy. In response, the Cuban government initiated radical improvements in energy efficiency, including a huge shift to localized, organic food production (Havana produces 50% of its own vegetables), solar energy projects, and mass car sharing. Cuba's GDP is only US$3,000 per person per year, but Cubans live just as long as people in the US, the same proportion read and write and fewer die in infancy.

In the wider-spread adverse conditions during the Second World War, miniature "victory gardens" produced 40% of America's vegetables, and Britons were urged to "dig for victory", in a successful (if temporary) drive toward self-sufficiency.

Communities in transition

In response to today's looming supply problems, there are now more than 100 official "transition" communities in 11 countries across the world, and their numbers are growing steadily. Their shared purpose is to convert to low-carbon economies – to show that humans can use as much creativity, ingenuity and adaptability on the way down the energy slope as we did on the way up.

High demand for scrap metals has inspired a rash of bizarre crimes, ranging from the stripping of lead from English church roofs to a spate of thefts of manhole covers in Shanghai.

As resources become ever scarcer, forward-thinking designers are considering their attitude to materials. Rejecting the old "take-make-waste" principle, they are striving for "borrow-make-remake" designs that "close the loop" of reuse.

The book *Cradle to Cradle* by William McDonough is the eco-designer's bible and the tome itself closes the loop: printed on recycled (and recyclable) plastic resins and inorganic fillers, it looks and feels like top-quality paper, but is wood-free.

POPULATION SATURATION?

The number of people on the planet is increasing by the equivalent of the entire population of Egypt every year and a population bigger than Europe every 10 years. By 2050 there'll be **9 billion** people on Earth. That's another China and the Indian subcontinent on top of the world's current population. How will we manage this growth? Will our new neighbours be an **unbearable burden** on an already overloaded planet, or can we **readjust** to live alongside them sustainably?

⇅ Full up

- An expanding population has led to **scarcity of key resources** that are under increasing, unsustainable pressure
- Without an arable revolution, more and more people will **go hungry** as the world's population continues to grow

- Rapid population expansion and change can increase the **likelihood of conflict** in fragile states
- There's no way the planet can support more people **aspiring to live as we do in the West**

THOMAS MALTHUS: A PRIMER

The English political economist Thomas Malthus (1766–1834) was one of the most influential and controversial theorists of population. He argued that population sizes are naturally self-limiting because at a certain point a society can no longer produce enough food to feed all its members. This limit is known as a "carrying capacity".

Contemporary experts still debate the Earth's carrying capacity, with varying estimates of how many people can live comfortably on our planet based on a wide range of environmental and social indicators. Some argue that we're already past the planet's carrying capacity.

Spaces available

- Sharing land equitably, trading fairly, eating a sustainable diet and generally **reducing our ecological footprint** are more important than simply slowing population growth
- In some rural areas a bigger population can **increase food production** to feed themselves without harming the environment

THE DEMOGRAPHIC WINDOW

The poorest countries tend to have the highest birth rates. Providing education and access to family planning strategies allows women in these countries to take control of their fertility and have fewer children. With a larger proportion of people of working age in relation to dependent children, productivity increases, more time and resources are available for education, and poverty decreases.

FOCUS: Population projections

Every year there are around 130 million births and 55 million deaths worldwide – so population is growing by around 75 million people a year.

Huge expansion is predicted in the developing world – the 50 least developed countries are set to more than double their populations by 2050. However, after 2050 it is believed that population will stop growing for the first time after 10,000 years of almost uninterrupted growth.

Some countries have shrinking populations. There are more deaths than births every year in Japan, while emigration is reducing populations in certain eastern European countries.

> Worldwide, there have been more births since the First World War than in the rest of human history.

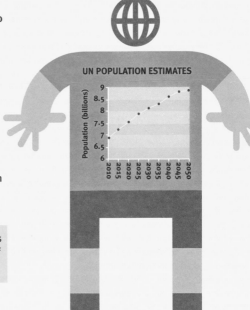

UN POPULATION ESTIMATES

Population (billions)

9
8.5
8
7.5
7
6.5
6

2010 2015 2020 2025 2030 2035 2040 2045 2050

How will we feed 9 billion hungry people?

The UN describes 37 nations as in "food crisis"; they have a combined population of 1.1 billion, and this will almost double to 2 billion by 2050. Globally, food production is increasing at a slower rate than population.

Some argue that genetically modified (GM) crops will play an increasingly significant role in feeding the world. GM crops are engineered to possess yield-increasing attributes, such as tolerance of herbicide, inbuilt resistance to pests, or ability to withstand hostile conditions. However, there are concerns over the long-term effects of GM crops on biodiversity and human health (for example, researchers have identified a potential link between GM soya and an increase in allergies to soya) – and critics also point to cases of GM crops that have lower yields than non-GM equivalents.

The world grain harvest tripled between 1950 and 2000, in large part due to a nine-fold increase in fertilizer use. But soil degradation and other environmental problems mean you can't keep adding fertilizer forever. So what's the answer? Agricultural adjustments such as "multicropping" (growing more than one crop a year on the same piece of land) and more efficient irrigation could improve yields, but no amount of innovation will increase yields indefinitely.

Sharing more equitably what we already produce may be a simpler solution – currently the overweight people on Earth outnumber the malnourished.

ONE CHILD ONLY

China's practice of restricting certain city-dwelling families to one child is estimated to have prevented 400 million births since the policy was introduced in 1979, but has had terrible unintended consequences. It is not uncommon for families to abandon or abort female babies, because their earning potential is lower than males. Not only has this caused untold suffering, but there is now a marked gender imbalance in the Chinese population – boys under the age of four outnumber girls by a ratio of 114:100.

Softer and safer measures, such as radio soap operas and programmes addressing family planning, have significantly reduced birth rates in places as diverse as Iran and Ethiopia.

Does more mean war?

Almost all the most unstable countries in the world, ranked on the Failed States Index, are experiencing severe "demographic pressure". It seems that in already unstable or "failing" states, having a steeply growing number of mouths to feed can increase the likelihood of conflict – particularly if the situation is compounded by a sudden environmental disaster. For example, the ongoing war in the Darfur region of Sudan is widely believed to have been triggered by severe drought.

Populations that are growing quickly have many more young people than old. At the end of the 20th century nearly 90% of countries with very young populations had unstable or totalitarian governance.

IT'S NOT HOW MANY LIVE BUT HOW THEY LIVE

While high-income countries contain 15% of the global population, they're responsible for 36% of the world's ecological footprint. Over the last 50 years, per capita ecological footprints have increased in the highest-income countries, where populations have stayed relatively stable, and decreased in the lowest-income countries, where populations continue to increase.
Source: WWF

▶ FOCUS: What will the world be like in 2050?

- **Urban** Half the world's population now live in towns and cities. By 2050, two thirds of us will be city-dwellers.
- **Looking for a date** There are 163 million more men than women in Asia and more boys born every year than girls.
- **More diverse** There are now 200 million migrants worldwide, who have mostly settled in North America and Europe. Climate change is likely to create yet more refugees over the next decades.
- **In need of another planet** If everyone lived as the average Westerner does, we'd need at least three planets to sustain us. If greener ways of living don't become the norm, by the 2030s the world's population could have a footprint twice as large as the Earth can support.

WHAT ABOUT CHINA?

In China a new coal-fired power plant opens every four days or so, and meanwhile India's greenhouse-gas emissions have risen by 83% since 1990. In the face of such escalating growth and pollution in developing countries, are our green efforts futile? Or should the West be setting an example by demonstrating that a high quality of life can be achieved through a low-carbon economy?

⬇⬆ Developing countries are wreaking havoc

- People in China, India and other developing countries aspire to the Western lifestyle – car ownership and consumerism are rocketing
- Some argue that countries tend to tackle environmental issues only when they've reached a certain level of wealth
- Developing countries' greenhouse-gas emissions are rising dramatically: China will emit more CO_2 in eight months than the EU aims to cut between now and 2020
- India and China's economies are heavily reliant on resource-intensive manufacturing

⬇⬆ Is the West guilty of double standards?

- Western countries have emitted 80% of all man-made CO_2 since the Industrial Revolution
- It's hypocritical to criticize developing countries for adopting the CO_2-emitting practices that we pioneered
- We've effectively exported our manufacturing emissions to the developing world: around a third of China's carbon footprint now comes from producing goods for export to the West
- The West not only has the greatest burden of responsibility, but is in the best place to lead change to a low-carbon economy
- There are huge economic opportunities in green technologies and knowledge already being taken up by India and China

Contraction and convergence

We need to stabilize and then reduce CO_2 emissions as soon as possible. But Western countries have had several hundred years of (very polluting) development, so it seems unfair to ask developing countries to cap their emissions at very low levels just because ours have got so high.

A possible solution is to agree that each country is entitled to an equal volume of emissions per person, based on what we understand to be the Earth's carrying capacity. The more polluting nations would have to reduce their emissions. Countries that used less than their allocation could sell spare CO_2 units to countries that exceeded their target, but would have to invest the revenue in zero-carbon technologies.

This system sets a safe and stable target for concentrations of greenhouse gases in the atmosphere, and a date by which those concentrations should be achieved, based on the best scientific evidence.

REASONS TO BE CHEERFUL

While it continues to build coal-fired power plants, China is making dramatic strides in renewable energy. In 2007 its investment in renewables was exceeded only by Germany. It is a leading manufacturer of solar photovoltaic technology and will become the world's leading wind turbine exporter by 2009.

Again looking on the bright side, although dramatic rises in car ownership are an undeniable concern, Chinese fuel-efficiency standards for vehicles are 40% higher than those in the US.

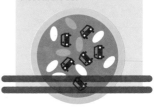

OUR FOREFATHERS' FOOTPRINTS

US: 20

UK: 12

China: 4.2

India: 1.2

Per capita annual CO_2 emissions (in tonnes)

China now emits more CO_2 than any other country, but its per capita annual emissions are still less than a quarter those of the US (see left). What's more, China has been a significant contributor to global warming for only 30 years, while developed countries have been belching out CO_2 for 200 years. So Western countries have made a far greater overall contribution to man-made emissions: in the UK and US the cumulative per capita carbon footprint is around 1,100 tonnes, compared with 66 tonnes for China and 23 tonnes for India.

NO TIME TO WASTE

We've become used to throwing things away, often after a single use – a concept that doesn't exist in nature, where **faeces are fertilizer** and **dead bodies are dinner.** Dealing with the billions of tonnes of stuff we discard each year has become a scourge of modern life. Is getting to grips with the **three Rs** – reduction, reuse and recycling – the only solution, or **can we put our waste to use?**

⬆⬇ A problem that's piling up

- The developed world is running out of suitable landfill locations and even the best sites are responsible for releasing **methane** into the air and leaching harmful **pollutants** into the soil
- Rapidly developing economies such as China and India are adding to the problem: it's predicted that by 2020 China alone will produce **as much waste** as the whole world did in 1997
- As well as creating waste, mass consumption **depletes** scarce resources
- The millions of tonnes that slip through the waste-processing net get dispersed all over the world, polluting the **soil, sea and skies**

THE GREAT PACIFIC GARBAGE PATCH

In OECD countries waste production has increased by 60% per person since 1980. And we're increasingly turning to single-use items made out of plastic, which can take centuries to decompose.

Sail 800km (500 miles) west of California and you'll hit millions of tonnes of such refuse bobbing over an area twice the size of Texas. Most of this toxic trash "continent" is plastic, which is partially broken down by sunlight, but never disappears. The UN estimates that every square mile of the world's oceans contains 46,000 pieces of floating plastic, which entangles and chokes more than a million seabirds every year, as well as over 100,000 marine mammals. In some ocean areas, plastic outweighs plankton by a ratio of six to one.

⬆ Can we put our waste to good use?

- It's possible to use waste as a source of energy, thereby **reducing direct dependence** on fossil fuels
- **Incinerating waste** reduces the amount that needs to be sent to landfill, and can produce affordable energy

- Another method of producing energy from waste, **anaerobic digestion** (see page 35) has been recognized by the United Nations Development Programme as one of the most useful decentralized sources of energy supply

---

RE-TYRE OR RETREAD

Each year about a billion tyres are sold worldwide, and in the US alone 300 million are thrown away annually – one for every man, woman and child. They're bulky, a potential fire hazard, contain valuable materials and toxic substances and break down very slowly in landfills.

Waste tyres can be incinerated to produce electricity. The largest tyre dump in the USA generates power for nearly 15,000 homes in this way. But incinerating tyres produces high levels of atmospheric pollution. A better solution is to retread rather than replace worn tyres. Retreaded tyres are as highly regulated as brand-new tyres, require half as much energy, and save the 25l (5.5 gal) of crude oil that goes into a new tyre.

FOCUS: The pros and cons of recycling

Recycling is much more efficient than throwing things away and starting again from scratch (see right), but still requires significant energy. And the circle's not endless – paper, for example, can only be recycled about seven times. The key to the waste challenge is very simple: we need to reduce the amount of waste we produce, and think of ways to reuse products rather than automatically discarding them.

Benefits of recycling over manufacturing from scratch	aluminium	steel	paper	glass
Reduction in energy use	90–97%	47–74%	23–74%	4–32%
Reduction in air pollution	95%	85%	74%	20%

Digging deep

Most of the world's refuse is sent to landfills – clay or plastic-lined holes in the ground. Burying rubbish is contentious. Sealed in without oxygen, waste breaks down slowly, releasing methane, a greenhouse gas 21 times more potent than CO_2. Poorly lined landfills can also leak polluting "leachate" into local groundwater. Suitable sites are hard to find, meaning that many cities have to transport their waste long distances. One landfill proponent calculated that a hole 47km (18 miles) square and 30m (100ft) deep could take 100 years' worth of US waste, but it's hard to imagine anyone wanting that on their doorstep.

Some countries, such as Germany, Austria and Switzerland, have banned the depositing of untreated waste in landfill – only ash from incinerated waste or the solid residue from anaerobic digestion (see opposite) is now acceptable.

A TAXING QUESTION

One way to restrict the amount of waste that is sent to landfill is to impose a landfill tax, as in the UK and in California. In the UK the landfill tax has been a key factor in encouraging local authorities to step up their recycling provision.

Some UK authorities have adopted additional measures to discourage excessive household waste generation, such as reducing the frequency of general waste collections and fining households who refuse to recycle. However, these policies are unpopular, and are believed to have contributed to recent increases in "fly-tipping" – unauthorized dumping of waste in alleyways or rural locations.

▶ FOCUS: One step forward, two steps back?

Because it has insufficient recycling facilities, the UK ships boatloads of waste to China to recycle – as much as 237,753 tonnes of plastic alone in 2005. This isn't as daft as it seems: since the UK imports far more goods than it exports, many ships – which need ballast to float properly – would have to carry water if waste wasn't available to load their return trip.

Going for the burn

Much of our waste can be burned, producing low volumes of ash, which can be used in the construction and road building industries (admittedly not the greenest of reincarnations). Some incinerators offer the added benefit of driving turbines to create electricity.

However, burning waste has its downsides. It produces CO_2 and dioxins (which have been linked to cancer), as well as environmentally-damaging arsenic and lead. When we incinerate waste, we use the resources it contains only once, rather than recycling them endlessly as would happen through natural processes. What's more, some environmentalists argue that using waste as fuel can actually encourage additional waste production: if people think that the waste they generate is going to be put to good use, they may feel less inclined to reduce their consumption.

PLASTIC BAGS

In many countries, plastic bags have become a potent symbol of outrageous wastefulness. While only the tip of the iceberg, these flimsy bags aren't insignificant: worldwide over a million are used every minute – most only once, for about 12 minutes. Nobody knows exactly how long the bags will take to degrade. Some estimate up to 1,000 years; some say they'll never completely break down.

▶ FOCUS: Anaerobic digestion

When organic waste breaks down in landfill, the methane and CO_2 produced is released directly into the atmosphere. However, if biodegradable materials are separated and put into an anaerobic digester, the gas (known as biogas) can be used as a fuel for electricity, heating or transport. In Sweden large-scale biogas plants provide fuel for urban bus fleets. As with incineration (see above), the major drawback of this technology is that it uses the waste only once.

WATER

Since 1950 the human population has more than **doubled** and its water use has **tripled**, while the amount of available freshwater has stayed almost the same. Ensuring everyone on the planet has enough clean, safe drinking water is one of the **greatest challenges** we face.

Is desalination the answer?

- Desalination has considerably reduced the need for groundwater extraction in some regions
- Saltwater is unlikely ever to run dry, and desalination should be **immune** to climatic changes
- **Emerging technologies** are helping to minimize the environmental impacts of desalination plants

Don't waste a drop

- As well as being decades away from significantly increasing water supply, desalination is very **energy-intensive** and has major impacts on marine ecosystems
- According to the UN there's enough accessible freshwater to meet demand – it's **mismanagement** that causes shortages
- If we all **use less** water it leaves more to go around, while reducing the need for energy-intensive processing
- Increasing the efficiency of extracting, transporting and using water are **cheaper and more reliable** ways of safeguarding water supply than desalination

Recycling water

"Grey" water from baths, showers and washbasins can be recycled so it is clean enough to use again. The process can be mechanical – passing water through a filter, disinfecting it and pumping it back to a house – or natural, with reed beds doing the job cheaply, and requiring little maintenance.

The growth of thirst

As a result of population growth, climate change, urbanization and falling underground reserves, by 2030 two thirds of countries could be "water-stressed", with demand outstripping supply. The likely consequences include an increase in water-borne diseases and heightened political tensions over water supplies.

Climate change and water

Las Vegas depends on Lake Mead (the largest man-made reservoir in the US), for its freshwater. The surface of the lake has dropped almost 30m (100ft) since 1999. Experts estimate there's a 50% chance it will run dry by 2021 if demand increases and rainfall decreases at current rates.

Relieving water stress

Some neat technologies are making a big difference in water-stressed countries. Lifestraw® is a short, lightweight pump that makes polluted water drinkable. Handed out in African countries, it can be worn round the neck, removes 99.99% of parasites and bacteria and lasts for a year.

Desalination: still a drop in the ocean

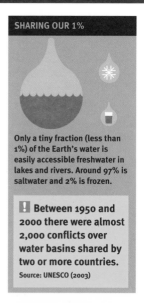

SHARING OUR 1%

Only a tiny fraction (less than 1%) of the Earth's water is easily accessible freshwater in lakes and rivers. Around 97% is saltwater and 2% is frozen.

> ❗ **Between 1950 and 2000 there were almost 2,000 conflicts over water basins shared by two or more countries.**
> Source: UNESCO (2003)

The two main ways of taking the salt out of saltwater are thermal distillation – heating to produce pure water vapour which is then cooled – and reverse osmosis, which involves forcing seawater through a series of membranes to separate out salt, minerals and marine organisms. Reverse osmosis is more commonly used – it requires less energy and the membranes used are becoming increasingly effective.

Approximately 15% of Israeli households get their fresh water via desalination – the highest proportion of any country. Worldwide, however, only 0.3% of water for human use is sourced through desalination. Plants are expensive and take a long time to get up and running, so it's unlikely that this figure will increase significantly until late this century. They're also very energy-intensive. Each year a medium-sized plant uses around 50 million kWh – enough energy to power around 10,000 homes.

▶ FOCUS: Water footprints

The water we use to flush our toilets and brush our teeth is only a small part of our water footprint. While each person in the West directly uses anything up to 35ol (77 gal) per day, the water used to produce the food, textiles and other products we rely on is many times that (see the chart, right, for some examples). When our direct and hidden footprints are added up, each person soaks up a staggering 4,000–5,000l (880–1,100 gal) per day.

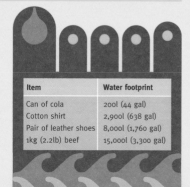

Item	Water footprint
Can of cola	200l (44 gal)
Cotton shirt	2,900l (638 gal)
Pair of leather shoes	8,000l (1,760 gal)
1kg (2.2lb) beef	15,000l (3,300 gal)

Repairing the damage

Enlightened policies in two of the world's major cities could provide blueprints for water management in the future:

Restoring the Catskills

Rainwater is naturally purified by forested areas, from which it runs into streams and lakes. For generations, New York City got exceptionally clean water this way from the Catskill Mountains. As the region's population grew, more and more forest was converted into farms, homes and resorts. These polluted the water until it was no longer drinkable. New York had a choice: spend $6–8 billion on a mechanical filtration plant that would cost $300 million a year to run, or $1 billion to restore the Catskills, with minimal maintenance costs. The answer was obvious.

Tale of the Thames

In 1215 the Magna Carta included laws to help fish swim safely through the Thames to their spawning grounds, but by the 1800s, London was growing rapidly and the river was used as the city's toilet. Perhaps unsurprisingly, there were five cholera epidemics between 1830 and 1871. 1858 was known as the Year of the Great Stink – some days parliament had to be abandoned because of the stench coming from the river. But new laws and improvements in sewerage meant the river got cleaner – so much so that in 1974 a salmon was caught from the Thames for the first time in more than 140 years.

WATER GOALS

Worldwide, only half of us have clean water on tap. Billions must walk miles to fill buckets from muddy water sources. One of the Millennium Development Goals is to halve the proportion of people without sustainable access to safe drinking water and sanitation by 2015.

❗ A lifesaver in water-stressed regions, the Aquaduct bicycle draws up and filters 20 gallons of water in its tank while a rider pedals. It can provide a whole family with enough freshwater to meet its needs.

CONSERVATION

Ninety-nine percent of the species that have ever lived have gone extinct, in a gradual process of evolution in which only the fittest survive. Since the Industrial Revolution, the rate of extinction has increased, leading to fears that a man-made "great extinction" is irreversibly damaging the ecosystems we depend on. But with so many species still out there, do we really need to worry that some are disappearing?

Another "great extinction"?

- Human activity, including rainforest destruction for timber and farming, pollution of habitats and unsustainable agriculture practices is irreparably damaging ecosystems
- A UN study has found that 10–30% of all described species are under serious threat of extinction

- Species are disappearing 1,000–10,000 times faster than historical extinction rates, leaving no time for evolution of other species
- The "services" provided by ecosystems – from clean water and air to food, medicines and shelter – are essential for life as we know it

THE PLIGHT OF THE HUMBLE BEE

Albert Einstein wrote "If the bee disappeared ..., man would have only four years of life left." At least a third of the food we eat depends on the pollinating power of the honeybee. The past few years have seen a dramatic drop in bee populations, known as "Colony Collapse Disorder". In the US, where bees pollinate $15 billion-worth of crops a year, 800,000 hives were wiped out in 2007 alone. Demand for commercial bees is increasing as supply dwindles – one Florida beekeeper transports his hives to California each year to pollinate the almond harvest.

↑ Can human ingenuity replace evolution?

- Optimists confident in the ability of scientists to come up with ingenious solutions believe that we'll be able to save sufficient species to **avoid catastrophe**
- Some anti-poverty campaigners argue that starving peoples cannot be blamed for prioritizing **food** production over **conservation** (but cropland gained by "slash and burn" is prone to soil erosion, and so the practice is **unsustainable**)
- Advances in genetics may enable us to bring species **back from the dead** in the future
- Environmental organizations are **protecting land** for endangered species to inhabit

AXES OF EVIL

More land was converted to cropland between 1950 and 1980 than in the 150 years from 1700 to 1850.

Source: World Resources Institute, Millennium Ecosystem Assessment (2005)

▶ FOCUS: Saving for later

Some of the world's animal and plant species are preserved in gene banks such as Norway's Svalbard Global Seed Vault. In theory, a species stored in a gene bank is protected against extinction. However, the technology does have its limitations. For example, the eggs and sperm of an animal, once thawed and fertilized, still need to be carried in a live female. And once an ecosystem has been destroyed, it's hard to imagine being able to reintroduce all the species that lived in it in the correct proportions to restore its natural balance.

Where to protect?

Biodiversity (the number and variety of different species in an area) varies from place to place, and experts disagree over conservation priorities. Over half the world's species live in the moist tropical forests which cover just 6% of the world's surface, so many environmental scientists see preserving such "hot spots" as a priority. As well as hosting charismatic "megafauna" like tigers and leopards, these regions provide vital but less tangible services, such as purifying the air and regulating climate.

Others argue that we should focus on protecting areas rich in species with particular uses to humans. For example, the Madagascar periwinkle, an active ingredient in a medicine that dramatically increases the chances of survival for children with leukaemia, is now extinct in its original habitat, the Madagascan rainforest, as a result of deforestation. It only survives because it was transported to other parts of the world once its healing benefits had been identified.

However we prioritize their protection, a diversity of species is key – not least because we don't know what will be precious to us in the future. If more than two types of potato had been available to Irish farmers, the terrible potato famine of the 1840s might have been averted, as some other types would probably have been more resistant to blight. As we head into a period of unpredictable climate change, who knows which crops and other species will survive where?

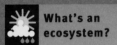

What's an ecosystem?

An ecosystem is a community of living things – plants, animals (including people), bacteria, fungi – that depend on each other and their physical environment (air, water, soil, sunlight) for survival. Every species has a part to play in this interactive web of life.

Threats to the ecosystem

Until now, krill have not been fished, because they release self-destructing enzymes when caught. But a trawler is in development that can pipe up and freeze the crustaceans before they deploy this mechanism, opening the way for mass exploitation of krill as a feedstuff in aquaculture. An additional threat comes from the increasing acidification of the ocean as a result of increased CO_2 levels, which interferes in the production of calcium carbonate, which krill (and other crustaceans) need to form healthy shells.

 ## The importance of krill

In the Southern Ocean ecosystem almost all life is supported by krill, a tiny shrimp that constitutes the greatest populational biomass of any multi-cellular species in modern evolution. Feeding on phytoplankton and zooplankton, this "keystone" species in turn feeds everything from blue whales and albatross to seals and dolphins.

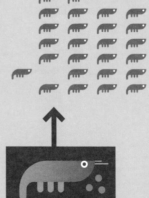

CHAPTER 2

THE WAY WE LIVE

Human ingenuity has given many of us lifestyles previously enjoyed only by the wealthiest few, but our modern way of life has evolved without much thought for the planet it ultimately depends upon. As populations grow and resources dwindle, the race is on to find ways of living that give as many people as possible a high quality of life while respecting and safeguarding the natural world.

This chapter looks at where and how we live, and asks:
- How do we make towns and cities greener places to live?
- Can there be such a thing as a low-carbon home?
- Can a profitable business be beneficial to the planet?
- What's the ecological footprint of fun?

CITY LIMITS

Today **half the human population** lives in towns and cities, and the proportion is still rising. **Clogged with cars** and scarred by layer after layer of redevelopment, modern cities present a host of **environmental challenges.** But could a shift to high-density urban living actually be good for the planet – the only **sustainable response** to a growing population? What are the most forward-thinking cities around the world doing to make themselves **greener**?

⬆⬇ Can urban areas be sustainable?

- Compact neighbourhoods with amenities close by and good public transport **reduce the need for cars**
- Urban dwellers often have **lower ecological footprints** than those in the country
- Well-planned new cities can be designed to make it easier for their inhabitants to adopt **sustainable lifestyles**
- More concentrated development in smaller areas can **free up land for farming and wildlife**

GETTING SOME AIR

London's particulate pollution (the main cause of pollution-related mortality) has fallen 22-fold since the late 19th century. In Paris, levels have fallen 66% since 1970, and in Athens by 43% since 1985. Los Angeles remains polluted but levels have been cut by nearly a third in the last decade.

Sources: Quality of Urban Air Review Group/OECD

ALL EYES TO THE EAST

Planned to cover an area a third the size of Manhattan and house 500,000 people by the time of its scheduled completion in 2040, the new city of Dongtan on an island near Shanghai aims to be carless, and completely self-sufficient in water and energy, with energy-efficient buildings. However, construction has not yet started and the project is in danger of slipping behind schedule.

⬆️⬇️ Or is city living bad for the planet?

- It's more difficult to be self-sufficient in the city than in the countryside
- **Air pollution** is a persistent problem in urban areas
- Without careful planning, cities can sprawl in a way that encourages **car dependency** and therefore carbon emissions

- As urban areas grow and develop, they rely on natural resources from further and further away to support them, which adds to carbon emissions
- City living can cut urban residents off from the natural world, meaning that the **impacts** of their lifestyles are often **invisible** to them

▶ FOCUS: All-consuming cities

Urban areas take up just 2% of the Earth's land mass but account for about 75% of industrial wood use. Similarly, 60% of the water withdrawn for human use goes to urban areas.

Taking London as an example, the "ecological footprint" of land required to support the consumption and waste disposal of the population of the city is around 120 times its surface area, or about 20 million hectares (49 million acres) – that's nearly equal to the productive land area of Great Britain as a whole.

Viewed in these terms, the ecological case against cities seems damning, but it is hard to imagine being able to accommodate the world's growing population without them. The challenge will be to plan and regulate urban development in ways that protect and enhance the environment.

❗ In 1950 New York was the world's only "megacity" (i.e. a metropolis with more than 10 million inhabitants). There are now 25, and 65 more cities accommodate more than five million people. The largest is the Greater Tokyo Area, with 35 million inhabitants – that's more than the whole of Canada.

NO URBAN MYTH

Only half the world's population lives in urban areas, but cities account for 75% of energy use and 75% of man-made greenhouse-gas emissions.

Source: Clinton Climate Initiative

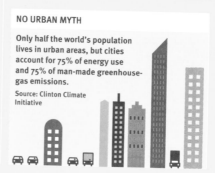

Where's the world's greenest city?

A growing number of places worldwide compete for the title of greenest city. Portland, Oregon tops the rankings in the US most regularly. The city generates half its energy from renewable sources, a quarter of its workforce commutes by bicycle, carpool or mass transit and it has 35 buildings certified by the US Green Building Council.

Arguably Scandinavia boasts the world's greenest cities. In Stockholm and Copenhagen, for example, a culture of environmental responsibility means people need no persuading to travel by public transport, or by foot or bicycle. An emphasis on shared space makes many Scandinavian cities efficient, compact places to live, and world-class architects compete to build green buildings in them.

GETTING OUT OF THE CAR

A 2005 Seattle study found that residents of neighbourhoods that contain a good mix of shops, houses and businesses, and have streets that are better connected (making walking and cycling easier and more convenient) travelled 26% fewer vehicle miles than residents of neighbourhoods that were more dispersed and less well connected.

Source: Lawrence Frank and Company, Inc.

▶ FOCUS: New green utopia?

Situated in the middle of the Abu Dhabi desert, the city of Masdar is being built from scratch using a combination of traditional methods (such as shaded walkways and narrow streets) and modern technologies (such as a mass transit system with personal pods on rails) to keep energy consumption low. Solar energy is plentiful in one of the hottest places in the world and the city aims to generate all its energy from renewable sources, send no waste to landfill, and recycle 80% of its water.

Starting from scratch makes it easier to avoid the legacy of high-carbon development that has to be dealt with in more established locations. But the middle of a desert is not the ideal location for a new development. And, like many other new eco-cities, Masdar is a long way away from existing settlements, meaning that residents' long-distance commutes could outweigh the city's environmental benefits.

A problem shared

Towns and cities worldwide are linking up to tackle environmental issues. The Large Cities Climate Leadership Group (C40) is a global network of 40 cities cooperating to reduce energy use and cut greenhouse-gas emissions. Through a set of shared goals and experiences, coordinated procurement of climate-friendly technologies and low-carbon activities they aim to help ease the transition to a low-carbon economy.

The ICLEI (International Council for Local Environmental Initiatives) partnership of over 800 city governments covers 300 million people in 68 countries. It helps towns, cities and even whole countries set and work toward environmental targets. It also campaigns for sustainable development through the UN.

! **While many cities are significantly improving their environmental performance, 1 billion people (18% of the world's population) still live in ramshackle, makeshift city slums. Often built on hillsides and flood-prone riverbanks from nothing more than cardboard boxes and corrugated iron, these settlements are acutely vulnerable to climatic changes.**
Source: UN Development Programme (2007)

PRODUCTIVE CITIES

Self-sufficiency can be nearly impossible for urban dwellers, but in a worldwide resurgence in urban food production an increasing number of townspeople are growing some of their food on their doorstep.

Caracas in Venezuela has 8,000 microgardens. The city aims to develop 100,000 such gardens, helping to reduce poverty through part-time jobs in farming, and improve health and food security.

Urban farming is also taking off in the US. In Chicago and Milwaukee, Growing Power teaches some of the most deprived communities how to create their own gardens. The organization also holds small markets so backyard growers can sell some of their produce for a profit. And on a larger scale, it has created a 1,900-m² (20,000-ft²) urban farm growing produce for soup kitchens.

Several projects in London are following a similar model. Food Up Front provides starter kits for people living on housing estates who want to grow something in their gardens, balconies, roof spaces, doorsteps or window boxes.

ECO-HOMES

What's the best way to reduce your home's carbon footprint? Do you need to make radical structural alterations – or even move to a brand-new "green house" – or can you make as big a difference by changing the way you live within your existing four walls?

⇅ Change where you live

- Even with the best intentions, it's almost impossible to minimize your footprint in an inefficient building
- Making buildings greener is a relatively easy way to ensure measurable impacts
- An ecological fit-out can help minimize a building's running costs, and can create a healthier place to live
- The economies of scale are making greener buildings increasingly affordable

⇅ Change how you live

- Buildings don't use energy, the people who use them do – the ways we use a building are often the most significant contributors to its environmental impact
- Even the most technologically advanced developments will have little effect on reducing the footprints of those who maintain environmentally damaging lifestyles
- The street layout, amenities and ambience of localities can be just as important as more ecological buildings in helping us live greener lifestyles

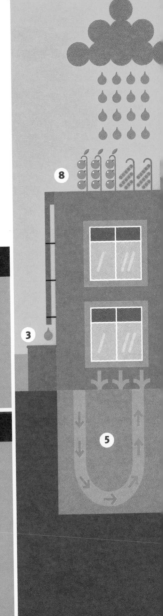

THE PERFECT GREEN HOUSE

To justify the energy and materials that go into its construction, a green building has to be built to last. It should use materials with low embodied energy (ideally recycled or reclaimed), that can easily be reused, and waste should be minimized during the construction process. The structure should be strong, resilient and adaptable to changes of use over time.

The following features (labelled on the diagram) are some of the most desirable.

1 Green roof planted in vegetation to absorb heat and reduce rainwater run-off

2 Reclaimed or Forestry Stewardship Council (FSC)-certified timber

3 Rainwater collection for irrigation

4 Greywater (water from dish-washing, laundry and bathing) recycling unit

5 Self-contained "power station" with integrated solar panels, wind turbine or, as shown opposite, ground-source heat pump

6 Wind-driven ventilation

7 Recharging point for electric car and/or bike store

8 Space for growing food (even if only window boxes or planters on balconies)

9 Passive heating, cooling and lighting – locating windows to make best use of sunlight and shade

10 Super-insulation and airtightness

Linking places and people

Urban design can be really important in making green living an appealing part of everyday life rather than a chore.

Good urban design focuses on making places where residents feel comfortable spending time outside. Dense, intelligently connected neighbourhoods enable people to cycle and walk without fear of aggressive car traffic. Attractive public spaces encourage people to meet up and enjoy outdoor activities, and a mixture of commercial and residential properties means people don't have to travel far to shop or work. Other aspects of good urban design include making space for local food production in allotments and gardens and encouraging independent retailers selling local produce.

GREEN LIFESTYLES OFFICER

Eco-developments often employ green concierges to advise on low-carbon living. In one study of the benefits of this strategy, the carbon savings were five to ten times greater compared to an equivalent investment in an on-site renewable energy source.

▶ FOCUS: The cost of green buildings

Housing developers might have you believe that green buildings cost a lot more to construct, but with economies of scale already kicking in, the price difference is narrowing all the time. Particularly when costs are considered over a building's whole lifespan, a low-carbon, ecologically sound building can be cheaper than a standard house.

The US Green Building Council estimates that the build cost of a house with their "LEED" (Leadership in Energy and Environmental Design) green certification is around 1–5% more than the cost of an equivalent uncertified one. Over 30 years, for an average-sized house, this amounts to a premium of up to $2 per day, but reduced running costs should outweigh the higher construction cost.

Source: US Green Building Council

Make do and mend

While state-of-the-art green buildings should be the template for development projects, we cannot tear down all existing housing stock and replace it with brand-new eco-homes. Not only would this be impractical but it would be a terrible waste of the materials that have gone into centuries of construction. Clearly, refurbishment has a key part to play. Taking the UK as an example, it has been calculated that it would cost around £13 billion per year to alter existing homes to minimize their carbon footprint. That sounds like a lot, but not when you compare it to the £23 billion already spent each year on general repair, maintenance and improvement of existing housing.

According to the Environmental Change Institute, the energy embodied in the refurbishment of an average house to make it low-carbon is about 15,000 kWh. The energy consumption of an average UK household is about 22,000 kWh per year, and advanced refurbishment could cut that by 50% or more. On this basis, the embodied energy of the refurbishment would be "earned out" in little more than a year.

Part of the green housing solution may be right under our noses – in the form of the empty, dilapidated houses that are a feature of every neighbourhood. Refurbishing and reusing an unoccupied property instead of building a new one is estimated to avoid the emission of 35 tonnes of CO_2 – almost twice the average American's annual carbon footprint.

THE GREEN PARADOX

According to one study, people who regularly recycle the most rubbish and save the most energy at home are also the most likely to take long-haul flights. Respondents said that doing good things at home made them less guilty about flying.
Source: Exeter University (2008)

❗ For the average person in the UK, CO_2 emissions break down as follows:
- **Energy consumed in the home 10%**
- **Personal transport 18%**
- **Embodied energy in the home 3%**
- **Waste and consumer items 13%**
- **Food 23%**
- **Shared services and infrastructure 33%**
Source: ESD

WORKING HOURS

In this era of heightened environmental consciousness, businesses of all kinds are keen to draw attention to their eco-credentials. Incorporating the sustainability agenda into successful business practice is undoubtedly a powerful force for change, but attempts by multinational corporations to present an ethical aspect to their operations often arouse suspicion. Is it really possible to do good business that benefits more than just the owners or investors?

⇕ Sustainability is good business

- The business case for sustainability has repeatedly been proven – it can increase **efficiency**, enhance **brand value** and open up whole **new markets**, leading to positive social, environmental and financial outcomes
- Businesses aren't faceless machines but groups of people who don't want to leave their values at home – ethical practices are increasingly important factors in **attracting and retaining good staff**
- The **unprecedented power** and reach of global corporations brings **unparalleled responsibilities**
- **Consumer pressure** on businesses to operate more ethically is at an all-time high and isn't going to go away

SHARE PRICE AND SUSTAINABILITY

A major study of more than 1,200 executives from a cross-section of industries and company sizes worldwide powerfully illustrated that a green philosophy can correlate with a successful business: the companies that rated their sustainability efforts most highly over three years saw annual profit increases of 9% and share price increases of 33% more than those who ranked themselves worst.

Source: Economist Intelligence Unit (2007)

⬆ Maybe greenbacks and green don't mix

- Some businesses have shown more ingenuity in **pretending to be green** than making real improvements
- The idea of **limitless business growth** is fundamentally at odds with **limited natural resources**

- Some currently necessary industries are **inherently unsustainable**
- Businesses are **not being sufficiently led by legislation and market pressure** to effect wholesale change across all sectors

---

A LOAD OF GREENWASH

Advertising regulators are having to crack down on a growing number of bogus green claims (commonly known as "greenwash") – ranging from oil companies claiming sustainability for switching to oil from tar sands (a more energy-intensive process than conventional oil extraction) to airports boasting of their carbon-neutrality without including aircraft emissions in their calculations. Companies should beware of using greenwash – the damage they incur from being named and shamed arguably outweighs the potential kudos they hope to gain.

🔲 In 2001 the US military announced it was introducing "green" bullets, replacing environmentally unfriendly lead (which pollutes soil and water) with tungsten and tin.

▶ FOCUS: Corporate Social Responsibility (CSR)

CSR managers or teams are now a common feature in larger companies. Designed to address environmental, social and other ethical issues, CSR activities range from one-off charitable donations to long-term strategic initiatives. Critics argue that creating a special department for this purpose detracts from the task of integrating sustainability right through a business. Defenders of CSR maintain that it is a valuable way for organizations with no previous experience of ethical operations to get this area of their business up and running. Ideally, after a short time, CSR should evolve into a stronger entity embedded in senior management, or become amalgamated into core business.

THE BOTTOM LINE

General Electric's Ecomagination products, ranging from low-energy lighting to hybrid trains, added US$12 billion to the company's bottom line in 2006. CEO Jeffrey Immelt said: "My environmental agenda is not about being trendy or moral. It's about accelerating economic growth."

Source: General Electric

Measuring success

Businesses have long used the lack of universally accepted measures as excuses for inaction on ethical issues. However, the UN-endorsed "triple bottom line" is gaining credibility. This expands the scope of accounting to rate factors such as carbon emissions and working conditions alongside turnover and profit margin.

Those who dismiss ethical accounting as an irrelevance should take note of the financial success of companies included in the FTSE4Good and Dow Jones Sustainability indices. However, until detailed ethical reporting is mandatory, these initiatives can only be part of the solution.

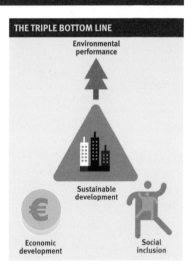

THE TRIPLE BOTTOM LINE

Environmental performance

Sustainable development

Economic development

Social inclusion

▶ FOCUS: Weaving a way to sustainability

Among the organizations that have done more than pay lip service to environmental concerns is Interface Carpets, the world's largest manufacturer of carpet tiles, which has set itself a target of becoming completely carbon neutral by 2020.

Among many sustainability initiatives, the company offers leasing services for floor coverings, takes carpet back to reuse and recycle, has pioneered natural, petrochemical-free glues and fibres and rewards employees for carbon-cutting ideas. Interface calculates that it has saved $372 million through waste reduction since 1995.

Can economic growth be sustainable?

Traditionally, the fundamental goals of business are to return a profit and achieve growth. However, can sustained growth be sustainable when it so often relies on increased consumption, which itself tends to entail increased energy use, resource and habitat depletion and waste generation? Isolated advances in efficiency and sustainability by pioneering companies can only go so far in curbing these environmental impacts. So, is legislation the answer?

The "polluter pays" principle

One instrument available to legislative bodies is "cap and trade". This involves a central government setting an overall cap on emissions and then dividing up this total into emissions allowances for individual businesses. If a company exceeds its allowance, it has to buy credits from one that has not used its full allocation.

In theory this creates a market in which heavy polluters are penalized and light polluters rewarded. However, there are drawbacks to the system. Judging by the example of the EU Emissions Trading Scheme (ETS), which started in 2005, it is difficult to set the allowances at the right level. In the first phase of the ETS, the calculations were so generous in some countries that few, if any, companies actually exceeded their allowance. To avoid this happening in the second phase of trading, it is proposed that emissions permits be allocated by auction rather than distributed freely.

LIGHTS OUT

In June 2007, an estimated 2 million light bulbs were turned off around London, showing how much energy is wasted by leaving lights on. In one hour, roughly 750MWh of electricity was saved – enough to run 3,000 televisions for a year. The Queen made sure Buckingham Palace joined in and the famous Piccadilly Circus advertising lights went out for the first time since the Second World War.

Source: Lights
Out London
partnership

■ The US Conference of Mayors estimated that "green jobs" could account for 10% of all new job growth in the US in the next three decades.

Source: Current and Potential Green Jobs in the US Economy (2008)

PLAYTIME

Everything we do has environmental impacts, and this includes the time
we spend relaxing, watching or playing sport, and going to concerts.
Some of the most obvious eco-offending sports, such as motor racing, are
under increasing pressure to get greener. Holding an Olympic Games
now brings the host country's eco-credentials under close scrutiny,
providing impetus for improvement. Is fun becoming guilt-free? Or are
leisure pursuits just a distraction from issues we can't afford to ignore?

⬇⬆ Game on for guilt-free fun

- Many major sporting events, from the **Super Bowl** to the **Olympics** and **Formula 1** motor racing, have developed wide-ranging **environmental programmes**

- High-profile events with a global reach can play a huge role in **raising awareness** and stimulating the **coordinated action** necessary to solve global environmental challenges

MUSIC MILES

The mammoth Live Earth concerts staged in 2007 were estimated to
have reached 2 billion people worldwide, and raised millions of dollars
for environmental causes. They were designed to be green both in their
message and their execution: 81% of the waste produced was diverted
from landfill, and, where possible, venues were low-energy and sourced
with green materials.

Even so, the musicians flew a total of 358,278km (222,624 miles)
(mostly by private jet) to perform at the concerts – nearly nine times the
circumference of the Earth. Including travel, it's estimated each concert
had a carbon footprint of 31,500 tonnes.

A record 15 million people watched the Live Earth concerts online. In
the future, the internet could avoid the need for flying at all. But would
virtual rocking around the world be such fun?

Game over for frivolous distractions

- Some sports, such as gas-guzzling Formula 1 or rally driving, seem **inherently** unsustainable
- Most of us pay little regard to the eco-credentials of our **"me time"** activities

- It's hard to cater for large numbers of people attending major sports events or concerts without resorting to **disposable** plastic cups and food packaging

AN ALL-CONSUMING PASSION

Sporting events can have enormous ecological footprints (the amount of land and sea needed to produce everything consumed) and energy use. The food and drink sold at the 2004 FA Cup final "used" 3,051 hectares (7,539 acres). Of course, the consumption isn't restricted to the stadiums: sales of televisions in the US increase by 60% during the week leading up to the Super Bowl, and a leading pizza delivery company reports a 30% increase in orders on the day.

Sources: Cardiff University ("Assessing the Environmental Consequences of Major Sporting Events: The 2004 FA Cup Final")/earth911.com

▶ FOCUS: Is Beijing better after the Olympics?

The 2008 Beijing Olympics aimed to be the greenest ever. Related environmental improvement projects included large-scale tree-planting and upgrading of public transport and sewage treatment plants. The "Bird's Nest" Olympic Stadium has a rainwater harvesting system to irrigate its infield and the "Water Cube" aquatic centre is wrapped in a high-efficiency thermal polymer membrane to minimize heating and cooling requirements.

However, Beijing still has some of the worst air pollution in the world. As a temporary fix, half the city's 3.3 million private cars were banned on alternate days during the Games, depending on whether their licence plates ended in an odd or an even number. But now they're all back on the roads in full force.

Source: UN Environment Programme (2008)

People power

Ever thought dancing was a waste of energy? Then you obviously haven't boogied on a dancefloor fitted with power-harvesting technology. A London nightclub has installed crystalline harvesters under its dancefloor to create tiny pulses of energy each time a dancer pushes down, helping to pay the venue's utility bills.

The same technology means that soldiers on the march, commuters pacing on train station platforms and even human heartbeats could be useful sources of energy. In Portland's Green Microgym, TVs, vending machines, lights and music run off electricity generated by the movement of people on exercise machines. At home, a Pedal-A-Watt cycling machine produces enough electricity to power a TV while you cycle.

 Football

An alternative is to work out without equipment. "Green gym" groups organize environmental enhancement projects such as hedge-laying, tree-planting and developing green public spaces. These give you the health benefits of outdoor exercise as well as helping your community. And it has been proven that connecting with the natural environment in this way increases feelings of well-being – a phenomenon known as biophilia.

The majority of the environmental impacts of most football games is a result of travel by the fans; simply shifting spectators from cars to buses could cut the overall impact of a typical match by at least a quarter.

Tickets to games at the 2006 football World Cup in Germany doubled as bus tickets, encouraging fans to travel more ecologically. The scheme was part of an ambitious "Green Goal" initiative, which also set high water and energy efficiency standards in the stadiums.

> ⚠ **A British inventor has devised a prototype for a washing machine that runs on pedal power. An hour's cycling generates enough electricity for one load.**

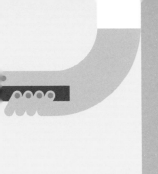

Golf

Golf courses need gallons of water to keep them green, and support little wildlife. Golf is also extremely vulnerable to changes in climate and the natural environment. According to one study, 69% of UK courses face serious threat from coastal erosion and/or flooding in the next 50 years.

However, an increasing number of golf clubs are striving to reduce their impact – for example, by harvesting rainwater for irrigation, leaving fairways to go brown in dry weather and planting weed-resistant grass varieties to minimize pesticide use.

Tennis

ennis is a relatively green sport. he courts are small, the stadiums re generally modest, and the ame itself is played with little nore than a racquet, ball and net.

The 2008 US Open marketed tself as a green event. The 2.4 nillion napkins used at the ournament were made from 90% ecycled paper; the 60,000 balls vere reused and then donated to ommunity youth programmes, and 00,000 free wallet cards printed vith eco-friendly tips were given to pectators.

Formula 1

In 2007, Formula 1 (F1)'s governing body banned investment in fossil fuel technologies. As a result, F1 teams' hefty R&D budgets should be ploughed into green technologies – some of which could end up in road vehicles.

Energy-regenerating devices (which save fuel by using energy created through braking) were introduced in 2009. Another proposal is that each team be set an energy quota that would be reduced each season, forcing teams into ever more ingenious energy-saving strategies.

CHAPTER 3

POWER TO THE PEOPLE

Human ingenuity has harnessed energy in previously unimaginable ways, and our appetite for power continues to grow. In the 200 years since its first sparks, electricity generation has become the world's single biggest consumer of fuel – mostly fossil fuel – and hence the world's largest source of carbon emissions. Now we face another challenge to our inventiveness: how to maintain our lifestyles as fossil-fuel supplies dwindle and climate change threatens.

This chapter examines the following questions:
- Is energy efficiency the way forward, or does our growing demand for power swallow any savings we might make?
- What is the future for fossil fuel?
- Does nuclear power have any place in a sustainable energy mix?
- Which renewable energy sources are a load of hot air and which hold water?

CAN LESS BE MORE?

As the human population keeps growing and finite resources get harder to come by, the race is on to find ways of maintaining a high quality of life while using less energy. Technological advances mean we can now generate energy and do countless tasks more efficiently than ever before. But these improvements are often outweighed by an overall increase in consumption. Can we rely on technology to minimize the impact of our energy-hungry lives? Or do we have to change our energy habits?

⬇⬆ The power of the efficiency drive

- Fossil fuel is finite, so it is vital that we get the most from remaining reserves, as well as using renewable and nuclear energy as efficiently as possible
- Power plants, cars, appliances, industrial processes and buildings all waste energy, whether through transmission losses along supply cables, heat escaping through poorly insulated structures or liberal use of the standby setting
- Wasting fuel is unnecessary, expensive and bad for the planet

IT ALL ADDS UP

Efforts by individuals to save energy may seem inconsequential, but added together the effect can be considerable. The equivalent of 64 million barrels of oil a day (almost 1.5 times current US energy consumption) could be saved worldwide through tried-and-tested efficiency measures – from upgrading building insulation to phasing out incandescent lighting. This would reduce the yearly growth in energy demand to less than 1% and provide around half the emissions reductions needed to stabilize greenhouse gases at what's thought to be a safe level.

Source: McKinsey (2008)

Efficiency's not enough

- Over the last 30 years, many products and services have become much more **energy-efficient**; but these laudable improvements have been **swamped** by more people using **more stuff** – meaning that energy use and CO_2 emissions are at an all-time high and still increasing

- The International Energy Agency predicts that worldwide we'll be using **30% more energy in 2020** than we use today
- It's not enough to make products and services more energy-efficient, we also need to **change the way we use them**

► FOCUS: The Khazzoom-Brookes postulate

Does increased energy efficiency reduce energy consumption? Not according to the economists Daniel Khazzoom and Len Brookes, who postulated that because increasing energy efficiency usually saves consumers money, this gives them more disposable income, which they more often than not spend on additional appliances which in turn use more power. Thus, perversely, increases in energy efficiency often actually increase overall energy use.

While the theory hasn't been proven beyond doubt, it holds true in a large number of contexts. For example, when higher-capacity passenger aircraft were introduced in the 1970s to meet growing demand, it was expected that the resulting increase in carrying efficiency would mean that fewer flights would be required. However, the opposite effect occurred – because the improved efficiency made air travel more affordable, demand soared.

STAND BY FOR THE TRUTH

Although many gadgets' chargers now consume little or no energy while they are plugged in but not connected up, standby can still be hugely wasteful.

Standby power consumption ranges from 20 to 60W, equivalent to 4% to 10% of total residential energy consumption. Yet the technology is available to reduce this to 1W and a global standard could mandate it. Better still, why not ban the setting entirely? A common misconception is that appliances use more energy powering up from "off" than from standby, but this is not the case for today's models, so standby serves no purpose.

Real-time pricing

When lots of people use energy at the same time – for example, running air-conditioning units on a hot summer afternoon – energy use spikes, so backup "peak-load" power stations must be fired up. These backup power stations tend to be the most polluting and least energy-efficient (which is why they aren't normally used), meaning that the environmental impact of energy spikes is even greater than the additional energy demand would suggest.

Real-time pricing (RTP) – charging consumers more for peak-time electricity – is a way to even out demand across the day so that inefficient backup power stations are not needed. The results of the New York City RTP pilot (see opposite) suggest that such a system could be successful. However, there are several obstacles and objections to be overcome before rolling out RTP on a large scale. For example, the "smart" meters required to measure electricity usage at different times of the day are at least three times as expensive as standard meters. And before making RTP compulsory, energy providers would have to find a way to protect from the most punishing rates low-income consumers who rely on electricity at peak periods.

> ⚠ A Japanese programme (Cool Biz) to minimize air conditioning use encourages office workers to dress more casually in hot weather. Over a summer, this saves enough energy to power a city of 250,000 people for a month.

THE "NEGAWATT" CONCEPT

Twenty years ago the energy expert Amory Lovins came up with the idea of the "negawatt" – a watt of energy that's left unused as a result of energy-efficient living, which could be accorded a value equivalent to the cost of generating a watt of energy. So, for example, an energy company might decide to invest in "negawatt production" by sending each of its customers a free energy-efficient lightbulb, rather than installing the additional generating capacity required to power the same number of traditional incandescent bulbs.

Legislation to ensure energy efficiency is being enacted worldwide. The US Energy Policy Act (2005) includes a rise in efficiency standards for appliances, which could reduce electricity demand enough to avoid building 29 coal-fired power plants.

Local energy production

Around 7% of the energy produced by large electricity plants is lost in transmission, so it can make sense to build smaller generators within communities. Combined Heat and Power (CHP) plants are particularly efficient because they capture the heat usually treated as a waste product of electricity generation. This can be used to warm houses or provide hot water. If powered by sustainably produced biomass (see page 82), CHP can be close to carbon neutral.

On an even smaller scale, some countries allow householders to sell any excess energy produced by solar panels or wind turbines on their homes to a larger grid at a premium price. This provides a huge incentive for individuals not only to invest in renewable energy, but to use it as efficiently as possible.

CHILLING STATS

Typically, the "green" thing to do is to make your possessions last as long as possible, but this principle doesn't apply to fridges and freezers. A modern, energy-efficient fridge uses half the energy of a pre-1993 equivalent.
Source: Energy Star

▶ FOCUS: New York real-time pricing pilot

As part of the PlaNYC plan to reduce New York City's CO_2 emissions, a small-scale, real-time pricing pilot was conducted between 2004 and 2006. The graph, right, shows that participants in the experiment responded to the variable price structure by reducing their energy consumption during the afternoon peak.

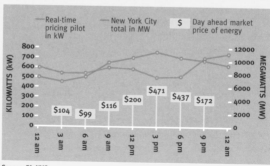

Source: PlaNYC

FOSSIL FUELS

It's hard to imagine what our lives would be like without fossil fuels, but the consequences of continuing to rely on them for the long term look too big for the planet to bear. Burning each barrel of oil or tonne of coal sends carbon that was gradually absorbed over thousands of years back into the atmosphere in an instant. Is this the end for fossil fuels or are there cleaner technologies around the corner?

⬇⬆ Fossil fuels are dead

- Fossil fuels are environmentally damaging to extract, burn and dispose of
- They are a finite resource, and as we use them ever faster supplies will inevitably dwindle
- Many countries are dependent on fossil-fuel imports from politically unstable regions, contributing to insecurity of supply and stoking political tensions
- Fossil-fuel prices underpin many countries' economies, and are fundamentally unpredictable

⬇⬆ Fossil fuels still have a future

- Carbon Capture and Storage could dramatically reduce emissions from coal-fired power plants
- Our energy needs are such that we are heavily reliant on fossil fuels for the foreseeable future

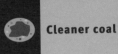 **Cleaner coal**

Another technology developed to mitigate the harmful effects of coal-fired power generation is IGCC (Integrated Gasification Combined Cycle). This involves converting solid coal to gas, then removing impurities such as sulphur dioxide, particulates and mercury before combusting the gas to generate electricity. Not only does this result in lower levels of pollutants, but it is more energy efficient than conventional coal power, and can be used in tandem with Carbon Capture and Storage.

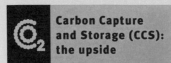

Carbon Capture and Storage (CCS): the upside

Realistically, the world is some years away from being able to abandon fossil-fuel power generation. CCS is an interim technology designed to reduce the harm caused by the CO_2 emitted in the process.

The idea is similar to a household extractor fan: CO_2 emissions are piped away from a power plant and carried deep into old gas or oil fields, or geological formations. In theory, the technology can reduce CO_2 emissions from power production by up to 90%. It also helps to dislodge the remaining, hard-to-reach deposits in oil fields nearing exhaustion.

Carbon Capture and Storage (CCS): the downside

CCS is still relatively untried and is potentially expensive, adding around US$1.5 billion to the cost of a new power plant. Without public subsidy, it's unlikely to be commercially viable for several decades, which is probably too late to avert climate change. And there are plenty of technical problems which need resolving, not least that if CO_2 escapes it turns the oceans acidic. Critics argue that we could develop large-scale, zero-carbon energy provision from renewable sources for the same amount of money that it would cost to get CCS off the ground.

Oil sands

Oil sands fields are a relatively new addition to the world's stated oil reserves. The process of extracting bitumen from the ground and converting it into oil or petroleum has only recently become cost-effective as a result of technological advances and increases in the market price of oil.

Oil sands have opened up a valuable new source of oil and the largest deposits are in Canada, a politically stable, benign regime. However, the drawbacks are considerable. The bitumen is obtained by strip-mining, a particularly damaging method of extraction. And oil from oil sands requires considerably more energy and water to extract and process than conventionally sourced oil.

LIVING ON A PRAYER

In 1900, the world produced 150 million barrels of oil. By 2000 production had grown to 28 billion barrels a year. In 2006, the world pumped 31 billion barrels of oil but discovered fewer than 9 billion barrels. Discoveries of conventional oil total roughly 2 trillion barrels, of which 1 trillion have been extracted so far.

Source: Lester R. Brown, *Plan B 3.0* (2008)

▶ FOCUS: Fossil fuels and the environment

When fossil fuels are burned, the carbon they contain combines with oxygen to form carbon dioxide (CO_2) gas. The quantity released depends on the carbon content of the fuel (see right). Burning fossil fuels is the single largest source of man-made atmospheric emissions, producing 98% of all US CO_2 emissions and around 80% of all US greenhouse gas.

	Coal	Natural gas	Oil
Relative CO_2 (per billion joules of energy)	24kg (53lb)	14kg (31lb)	20kg (44lb)
Other environmental effects	Coal mining destroys biodiversity and can contaminate groundwater. Burning coal leads to acid rain, and produces heavy metals and radioactive material.	Natural gas is generally cleaner than coal or oil, but still causes air pollution when burned, and its extraction affects biodiversity.	Spills from oil tankers can be devastating for marine and bird life. Biodiversity is affected by extraction at sea and on land. Oil combustion causes air pollution.

Source: United Nations University, Sustainable Environmental Futures programme (1993)

Energy security

The fossil-fuels debate is influenced by factors other than how much is left in the ground. Political, commercial and even terrorist machinations can leave importer countries feeling extremely vulnerable. Analysts are advising that to achieve long-term energy independence it is necessary to invest in renewables (see pages 76–83).

Geopolitics

Fossil fuels are inextricably linked to global politics. They've been key drivers and shapers of conflicts, ranging from the First World War to Iraq's invasion of Kuwait in 1990 and present-day tensions in the Middle East. Many pipelines run across national borders – a common source of unease. Ongoing wrangles over oil-extraction rights in the Arctic could end in political as well as ecological disaster.

Cartels

Oil-importing countries have to pay prices set by a number of powerful cartels. In 1973, OPEC raised oil prices by 60% in an instant, sparking a global energy crisis.

Global terrorism

The world economy depends on crucial supply lines, such as the trans-Alaskan oil pipeline and the Saudi export facility Ras Tanura which handles 6 million barrels a day. If these were disabled by an attack, individual countries' "strategic reserves" would on average last just 90 days.

OIL: IT GETS INTO EVERYTHING

Oil is incredibly useful as a raw material as well as a fuel. If you're sitting on a plastic chair it's likely to be made with petroleum, which is also in drinks bottles, carrier bags, insulation (from cables to sleeping bags), buckets, bowls, accelerator pedals in cars, glues, cling film, toys and even the quilting of your coat. It all adds up: the US uses more than 17 million barrels of oil each year just to satisfy its thirst for bottled water – enough to fuel a million cars for a year.

Source: Food and Water Watch (2007)

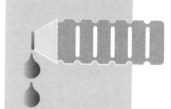

⚡ Six of the world's ten most profitable companies sell fossil fuels. Their collective carbon footprint is around ten times that of the other four combined.
Source: Fortune

NUCLEAR POWER

When a uranium atom is split apart it releases a huge amount of energy. In a nuclear power station, this energy is channelled into heating water to create steam which drives turbines that generate electricity. Low in CO_2 emissions, nuclear energy production is, according to some energy analysts, a crucial step in the transition to a low-carbon world. But others see no place for nuclear power, pointing to the human and ecological health risks associated with it – not to mention its expense.

⬇⬆ Nuclear is unavoidable

- Some argue that nuclear is the only high-volume, low-emission technology that can replace fossil fuels in the short term
- Nuclear power stations can supply a constant "base load" of electricity, which could complement any supply fluctuations from renewables
- Unlike some renewable energy sources such as hydroelectric and geothermal, the siting of nuclear plants isn't dependent on a particular physical geography
- Nuclear power is controlled by some of the strictest safety regulations of any major industry
- Modern nuclear power stations produce less waste than their predecessors
- Relatively large supplies of uranium are still available

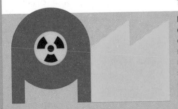

THE MYTH OF THREE MILE ISLAND

Nuclear power currently provides 20% of US electricity and 70% of the country's non-fossil fuel power, which suggests that it could play a role in meeting President Obama's stated aim of reducing US carbon emissions by 80% by 2050. Advocates of nuclear power maintain that no US nuclear facility has ever suffered a harmful malfunction. Even in the notorious Three Mile Island meltdown of 1979, they argue, radiation was successfully contained and there were no fatalities.

Nuclear is unthinkable

- The risk of radioactive contamination from a nuclear **meltdown** may be small but it's ever present and the consequences could be **catastrophic**
- There are still **no facilities** for the safe, long-term **storage** of radioactive waste
- Above-average incidences of childhood **leukaemia** have been found near nuclear reprocessing plants
- Even if new plants were built as quickly as possible, they wouldn't reduce carbon emissions quickly enough to avoid significant climate change
- Nuclear energy is based on a **centralized** system (a small number of large plants), which is a less efficient structure than a decentralized supply system (the basis for many renewable energy technologies)
- Certain **rogue states** may use their nuclear power industry to hide development of **nuclear weapons**
- Nuclear power has hitherto been feasible only with huge **state subsidies**
- Some argue that investment in nuclear power **diverts** both **finance** and **ingenuity** from renewable power and energy-efficiency programmes

> ! The IPCC predicts that sea levels will rise by at least 50cm (20in) by 2100, which would put the world's many coastal nuclear power stations at risk of flooding.

FOCUS: Does nuclear add up?

Some claim that nuclear power is the only major low-carbon source of electricity currently available to us. However, critics remain unconvinced, asking the following questions:

- **Is nuclear really low-carbon?** It is estimated that 14 million tonnes of concrete go into the construction of a nuclear power plant. A tonne of CO_2 is emitted for every tonne of Portland cement manufactured.
- **Is nuclear really ready to take over from fossil fuels?** Many of today's nuclear power plants will soon need to be decommissioned and it takes many years to build a new plant. Only a few companies in the world can produce the forgings able to withstand the extraordinary pressures inside a nuclear reactor.

WASTE OF TIME?

There is no permanent storage site for nuclear waste anywhere in the world. The US nuclear industry designated a mountain in Nevada for this purpose in 1978. So far $9 billion has been spent on the facility, which is due to open in 2017.
Source: US Department of Energy (2008)

A proliferation of fear

The international Nuclear Non-Proliferation Treaty gives member countries the right to run civilian nuclear energy programmes as long as they can demonstrate that these aren't being used for the development of nuclear weapons. It is theoretically possible to use a nuclear power facility to produce weapons-grade uranium or plutonium, although the plant would have to be run in such an unusual manner, with frequent shutdowns for refuelling, that it would be hard to cover up this activity from observers policing the treaty.

Non-signatory states are largely excluded from the trade in nuclear plant and materials. Despite this, a small number of countries, including India, Pakistan and Israel, are known or are believed to have developed unauthorized nuclear weapons.

Safety

The Chernobyl disaster of 1986 is a chilling reminder of what can happen when a nuclear power station fails. More than 300,000 people had to be resettled and the number of deaths attributable to the fallout could run into the thousands. Restrictions on foodstuffs will still be in place in the 2020s in some affected areas up to 3,200km (2,000 miles) from Chernobyl, because radioactive caesium from the fallout is lingering in the environment much longer than scientists anticipated.

However, modern plants are considerably safer. Some of them incorporate "walk-away" safety mechanisms to shut themselves down at the first sign of malfunction.

▶ FOCUS: Is fusion the future?

Nuclear fusion is the opposite process to fission: it involves forcing small atoms together to form larger atoms, releasing astonishing amounts of energy. Occurring naturally in stars and the sun, the process has been carried out under laboratory conditions, but is a long way from being used as a reliable means of generating energy.

If it were possible to harness this kind of power, it could be a dream energy source: waste-free, inherently safe and inexhaustible for the foreseeable future. Lake Geneva alone contains enough deuterium (a heavy form of hydrogen found in nature and used in nuclear fusion reactions) to satisfy the world's energy needs for several thousand years.

Ecology

Radiation alters ecosystems in serious and unpredictable ways. Radiation in rain accumulates up food chains, from soil to plants to animals, then potentially to humans. And if one animal or plant develops a genetic mutation as a result, the stability and diversity of the whole system – the biosphere on which we depend – is at risk.

Waste disposal

Radioactive waste will always be a major drawback of nuclear power, but power plants are 50% more efficient than they were 25 years ago and improvements are continuing to be made. Currently, waste is stored locally at the reactor sites. Permanent geologic storage facilities deep within rock are being developed by countries such as the US and Finland.

Global warming

Although nuclear power plants emit considerably less CO_2 than fossil-fuel equivalents, they do contribute to global warming in other ways, including the super-heating of nearby water sources used in reactor cooling systems. If nuclear power is to be considered at all, it should only be as a stop-gap technology to curb emissions until renewable energies are able to take the strain.

Cost

Compared with fossil fuels, nuclear energy is expensive. Opponents of nuclear argue that renewable energy would be ready for large-scale deployment much sooner if it received the level of funding given to the nuclear industry.

RENEWABLE ENERGY

As fossil fuels become more difficult and expensive to extract, restrictions on carbon emissions tighten, and doubts about nuclear persist, we are looking to renewable energy to play an increasingly prominent role in powering the 21st century. Are renewables ready to take the strain? What obstacles do they still need to overcome?

⬇️⬆️ Game on

- Once the infrastructure is built, renewable energy is **carbon free**
- **Operating costs** for renewable power generation are low
- Renewable energy enables countries with limited fossil-fuel resources to achieve **fuel independence**
- The technology for renewable energy is developing at a fast pace, improving efficiency and cost-effectiveness all the time

FULL STEAM AHEAD

Iceland is probably the world's most energy-advanced nation. It meets around 70% of its total energy requirements from renewables (mainly hydroelectric and geothermal) and generates 99.9% of its electricity from renewable energy. The country has been running a small fleet of hydrogen fuel cell buses since 2001 and the world's first commercial hydrogen refuelling station was opened in Reykjavik in 2003. Iceland hopes to be the first country to run entirely on renewable energy by 2050.

⬇️⬆️ Dream on

- Renewables are many billions of dollars and years away from providing a **significant** proportion of the world's energy needs
- Areas of extreme climate or geography often associated with high levels of renewable energy tend to be a long way away from the areas of highest demand
- While low in atmospheric pollution, renewable energy generation can be **visually polluting**, with many of the most suitable sites found in areas of outstanding natural beauty
- Many renewable energy sources are **intermittent and unpredictable**, making reliable provision a challenge

WIND

One of the most flexible renewable energy sources, wind can be harvested on land or at sea, by a large-scale wind farm or a single household. The basic technology is proven and turbine designs are becoming increasingly energy- and cost-efficient.

Offshore wind farms offer particular advantages: they're out of sight, benefit from higher wind speeds than their land-based cousins, and have little effect on marine life. In fact some studies suggest that the base of turbines host biodiverse, reef-like ecosystems.

Arguably the main drawback of wind power is the intermittent, unpredictable nature of wind supply (see right), which means that, at present, it is unsuitable as a provider of base-load electricity. And some people object to the visual impact of wind farms, which are often sited in particularly unspoilt, picturesque landscapes (although other people feel that wind turbines enhance the beauty of a landscape).

SUPERGRIDS

Two major problems with wind energy are that the wind doesn't blow constantly or predictably, or focus on areas with high energy consumption. It's a real challenge to store wind energy – options such as batteries, hydrogen fuel cells and underground compressed air are expensive, inefficient or untested.

Many experts believe that the answer lies in high-voltage direct current (HVDC) "supergrids", which can transport electricity from a variety of renewable energy sources over long distances with minimal loss in transmission. In the US, for example, the proposed Unified Smart Grid would allow wind energy from the northeast coast to power central regions and midday sun from Arizona to power the afternoon peak demand in the east.

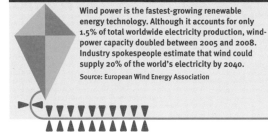

WHAT SHOULD WE EXPECT?

Wind power is the fastest-growing renewable energy technology. Although it accounts for only 1.5% of total worldwide electricity production, wind-power capacity doubled between 2005 and 2008. Industry spokespeople estimate that wind could supply 20% of the world's electricity by 2040.

Source: European Wind Energy Association

SOLAR

There are two main ways of harnessing the sun's energy. Photovoltaic (PV) cells convert sunlight into electricity. Light creates an electric field across one or two layers of a semiconducting material (usually polysilicon), causing electricity to flow. The more intense the light, the greater the flow of electricity. In contrast, solar thermal cells absorb heat, which is stored as hot water or gas in an insulated container and used for space heating and hot water.

Worldwide, only a small proportion of our electricity is generated by photovoltaic cells, but capacity is increasing by around 40% each year. Around 40 million homes around the world get hot water from solar thermal cells. Most of them are in China (see right), but their numbers are due to increase rapidly in countries such as Spain and Israel that have introduced legislation to incentivize or enforce the installation of solar panels on new houses.

One of the major advantages of solar photovoltaic power is its ability to bring electricity to remote communities that are not linked to a supply grid – an extreme example being space stations and satellites. Closer to home, solar is in prime position to become the electricity provider of choice for hot, sunny countries in the developing world.

Solar panels are also very space-efficient and unobtrusive – they can often be sited on a south-facing roof or wall – and require minimal maintenance.

SING IT FROM THE ROOFTOPS

Solar electric and solar thermal industries are booming in China. More than 2,000 Chinese companies manufacture solar water heaters, bringing hot water to many villages for the first time. China's solar installations harness as much energy as generated by 54 coal-fired power plants.

However, there is a trade-off for this rapid expansion. The process of converting the common semiconductor polysilicon into a usable form produces the highly toxic waste substance silicon tetrachloride. To meet demand and to avoid the expense of reprocessing, unscrupulous manufacturers have been dumping it and endangering human health.

⚠ **Despite increasing investment, worldwide solar power generation currently adds up to less than a third of the output of China's Three Gorges dam.**

Although, of course, they will not provide any energy at night time, modern solar panels can still be moderately effective even on cloudy days.

The shady side of solar

Some of the materials used as semiconductors on photovoltaic panels are scarce and expensive; others, such as polysilicon, require a huge amount of energy to convert into a usable form. This means that the startup costs of a solar electric power installation are high. Another drawback is that high levels of air pollution can affect efficiency, which may make solar relatively unsuitable for city-centre locations.

INSPIRED OR MISGUIDED?

Germany has invested so heavily in the installation of photovoltaic panels that it now generates more than half the world's solar electricity. Critics argue that this money would have been better spent on solar electricity provision in a more suitable country – such as Spain, which enjoys twice as much sunshine as Germany.

⚠ Every day, 8,000 times more energy than we currently use reaches us from the sun.

◀ FOCUS: The power of mirrors

Concentrated Solar Power (CSP) uses mirrors to concentrate the sun's heat to create steam to drive an engine or produce electricity. This tried-and-tested technology works best in the world's hottest places, where it's starting to be used on a large scale. The World Bank is helping to fund projects in Egypt, Mexico and Morocco, and other projects are in development in Algeria, China, India and South Africa. Around 2,000MW of CSP capacity are planned or contracted for in the US.

The sun doesn't shine equally everywhere, but its power can be shared. Plans are being developed to supply Europe with power generated by CSP in North Africa, delivered via an HVDC supergrid (see page 77). It is hoped that by 2050 a small area of African desert could help to meet Europe's electricity needs, as well as providing local power.

SOLAR-SIZE ME

Meeting the entire electricity needs of the US via photovoltaic systems would require an area of about 34,000km² (13,100 mile²) – that's less than 0.4% of the country's land area.

Source: Weinberg and Williams (1998)

HYDROELECTRICITY

Hydroelectric dams create electricity by using a controlled downward flow of water to turn a turbine. They're the only source of renewable energy currently used on a large scale for electricity generation – approximately 20% of the world's electricity comes from hydroelectric sources.

Although they avoid the need for fossil fuels, some argue that dams aren't truly renewable (see below) because they often severely alter ecosystems and displace thousands of people worldwide each year.

Hydroelectric output can be highly variable. Worldwide, hydroelectric generation increased by 7% between 2005 and 2006, second only to the increase in wind power. But between 2006 and 2007, electricity produced from dams fell by 14% – primarily because of reduced rainfall and snowfall. As the climate changes, hydroelectricity could become less and less reliable.

SMALL IS BEAUTIFUL

One of the major drawbacks of hydropower is the profound impact that "mega-dams", such as the Three Gorges dam in China, the world's largest, have on the surrounding landscape and population. For this reason, there are very few remaining sites where it would be feasible to install a large-scale hydroelectric facility.

However, small-scale "micro-hydro" plants producing up to 100kW offer far more potential for expansion. They are a particularly good option for small, remote communities in the developing world, and work well in tandem with photovoltaic power generation, with the hydro plant taking the strain in the winter when rainfall is higher and the solar facility coming into its own in the summer when sunlight is at its most plentiful.

▶ FOCUS: Is hydroelectricity renewable?

Large-scale dams have a major impact on the environment: they affect migration routes of fish, animals and birds, reduce water flow downstream and flood the river banks and surrounding land upstream. The flooding of the land causes methane to be released. Add to this the CO_2 emissions from the manufacture of the tonnes of concrete that go into a "mega-dam" and it is easy to see why some experts criticize large-scale hydroelectric installations.

WAVE AND TIDAL

Energy from the sea's tides and waves is perhaps the least utilized of all renewable sources, but it offers great potential. A major advantage of tidal power in particular is that, unlike wind and sunshine, tides are predictable and perennial. It is not believed that wave or tidal energy plants have any serious environmental impacts – indeed tidal plants can help protect coastlines against storm surge tides.

Perhaps the biggest challenge to be overcome is in designing machines that can convert tidal or wave motion into electricity efficiently while withstanding the high pressures exerted by turbulent ocean waters. There are high hopes that the Pelamis system (see below) used in the world's first commercial wave farm, situated off the Portuguese coast and in operation since 2008, will achieve the goal of supplying cost-effective electricity and provide a model for other developments.

A RIVER RUNS THROUGH IT

A tried-and-tested means of harnessing tidal energy is by constructing a barrage across a bay or a tidal stretch of a river. The barrage contains turbines, which generate power when water flows in or out through them at high and low tides. The first barrage tidal generating facility, across the Rance river in northwestern France, was built more than 40 years ago and is still operating today.

❗ Ireland has the world's most ambitious wave power development goal: 500MW (electricity for around a million homes) by 2020, enough to supply 7% of its electricity.

FOCUS: How the Pelamis works

Taking its name from a surface-swimming sea snake, the Pelamis wave power system consists of a series of strings of articulated cylinders. Each string harnesses energy from the wave-induced motion (both up and down and side to side) of its joints. This motion is resisted by hydraulic rams which pump high-pressure fluid through hydraulic motors that drive electrical generators.

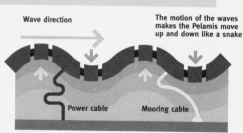

Wave direction

The motion of the waves makes the Pelamis move up and down like a snake

Power cable   Mooring cable

Source: Ocean Power Delivery Ltd

BIOMASS

Biomass is fresh organic matter that is used as fuel. If managed sustainably, biomass production is as close to carbon neutral as any fuel can be (see below). It can be an ideal way to use material that would otherwise go to waste, such as straw or by-products from the pulp and paper industries. Another factor in its favour is that marginal land that's unsuitable for food crops can be converted to biomass crops without loss of biodiversity.

However, biomass resources are finite, and when inappropriately used the impacts can be disastrous for ecosystems. As with biofuels (see pages 90 and 93), biomass crops can compete for land space with food crops. If land previously taking in carbon is cleared to plant them, and energy-intensive production processes are used, the CO_2 emissions from the total life-cycle of biomass crops can be high.

THE WORLD'S FUEL

Traditional biomass contributes 13% of all global final energy consumption. However, that figure is misleading. An estimated 2.5 billion people – nearly half the world's population – rely mainly or exclusively on biomass for their daily energy needs. They just happen to be people who use little or no electricity, using biomass for heating, cooking and hot water.

Source: Nick Middleton, *The Global Casino* (2008)

HOW IS BIOMASS CARBON NEUTRAL?

If plants are continually replanted, each new crop will take in the same amount of carbon over its lifetime as was released by burning the last crop, meaning that the process produces no net increase in CO_2. However, rainforest and other biomass-rich ecosystems can take millennia to develop, so they can't simply be replanted. Biomass supplies therefore have to be carefully chosen and managed.

⚠ The UK has an ambitious biomass target. If plans by the coal power operator Drax come to fruition, 10% of total UK electricity could be generated using biomass by 2014.

GEOTHERMAL

The earth a few feet beneath our feet stays at a constant temperature of between 10 and 16°C (50–61°F) almost everywhere in the world. Deeper down are layers of intensely hot rock and liquid. This heat can be used directly or converted to electricity.

Carbon emissions from geothermal energy can be less than 0.1% of those from fossil-fuel power plants and large-scale geothermal energy production is economically competitive with other forms of energy production. What's more, geothermal plants can operate continuously, whatever the weather, and the supply of heat from the ground is inexhaustible.

Individual homes can install their own geothermal energy systems (see below), although this can be expensive and require a relatively large area of land.

INTERESTING BORING

By boring to depths of over 3km (2 miles) below the Earth's surface, where steam is super-heated to temperatures reaching 450–600°C (840–1110°F), Iceland's energy experts hope to tap ten times as much energy as they currently harness. Icelandic geothermal energy could be one of the major power sources supplying a European supergrid, if a way can be found to connect Iceland to the European grid with minimal transmission losses.

🛈 One of the few environmental impacts of geothermal plants comes from hydrogen sulfide emissions which can impoverish lichen growth.

▶ FOCUS: On home soil

A domestic ground-source heat pump works a little like a refrigerator in reverse. Just as your fridge circulates refrigerant fluid which extracts heat from the inside of the appliance to keep your food cool, a heat pump circulates the same sort of fluid through a network of pipes in the ground to extract the earth's natural warmth and uses it to heat your home. In hot weather the process can be reversed to keep your home cool. The mechanism does require some fossil-fuel input to power the pump – unless, of course, your home is supplied by renewable electricity.

Fluid pumped through the ground absorbs the earth's natural warmth

CHAPTER 4

WORLD IN MOTION

Trains, planes and automobiles have made us more mobile than ever before. But with 3 billion cars anticipated on our already clogged roads by 2050 if we stick to our current transport strategies, and cheap flights making jetsetting a stroll in the park, are we travelling in the right direction?

This chapter investigates how we might turn the corner:

- Can we make aviation green enough to be sustainable or do we need to keep our feet on the ground?
- What will be the fuels of the future? How long will it be before hydrogen, renewable electricity and biofuels can take the strain?
- How can we make our cars as efficient as possible?
- Are we condemned to car craziness, or will gas-guzzlers be overtaken by high-speed trains and smart buses and undercut by low-carb cycling and walking?

AIR TRAVEL

Air travel has shrunk the world and opened up new horizons in business and tourism. But the phenomenal growth in jet-setting has come at a high environmental price. While other industries may cause more pollution, aviation has the fastest-growing emissions of any sector. As the skies become ever busier, can better timetabling and new, more fuel-efficient planes provide truly green air travel? Or are campaigners who object to proposed new runway sites doing us all a favour?

⬆⬇ Sky-high aviation growth needs to be curbed

- Aviation's CO_2 emissions are rising by around 5% each year – and their impacts are magnified because they occur high up in the atmosphere
- Planes release nitrogen oxide, which impacts on human and environmental health – studies in the US have identified a higher incidence of certain types of cancer around major airports
- Often built in the rural outskirts, airports wipe out biodiversity and produce significant light and sound pollution
- Much air travel is over distances that can be covered comfortably by rail or road
- Advances in communications technology mean business travel is no longer such a necessity

741 million passengers

1.2 billion passengers

FLIGHT NUMBERS

- The Federal Aviation Administration expects 1.2 billion passengers a year to travel on US carriers by 2020, up from 741 million in 2006
- The number of commercial jetliners is expected to double to 36,000 in the next 20 years
- 45% of flights within Europe cover less than 500km (350 miles)

Source: New Economics Foundation/World Development Movement

Aviation is getting greener

- Modern aircraft are up to 50% **more fuel-efficient** than their predecessors of 30 years ago
- Engine manufacturers are committed to **achieving reductions** in both CO_2 and nitrogen oxide emissions with each new generation of design

- Simple measures such as making **routes more direct** and changing **wing design** have the potential for modest emissions reductions
- Alternative energy sources such as **biofuels and solar** could reduce emissions over the coming decades

WHERE THE WIND BLOWS

Traditionally, aircraft follow set flightpaths so that air traffic controllers can safely monitor their progress. However, in 2005 the Australian Advanced Air Traffic Services introduced the Flex Tracks system, which allows pilots flying in Australian airspace to adjust their course to pick up the strongest tailwinds, and so reduce flight times and fuel consumption. In the first 50 days of operations between Singapore and Australia, the system cut CO_2 emissions by around 1,600 tonnes.

▶ FOCUS: The plane truth about CO_2

Ignoring the multiplying effects of "radiative forcing" (see page 88), aviation accounts for around 4% of man-made CO_2 emissions – equivalent to those produced by the whole population of Africa. What really marks the air industry out for attention is that it is the world's fastest-growing CO_2 producer, with emissions projected to increase by 50–70% by 2050. In this scenario, even if all other industries achieved the carbon reductions that they are aiming for, these savings would be outweighed by the increase in aviation emissions.

Sources: Tyndall Centre for Climate Change Research/New Economics Foundation/World Development Movement

Radiative forcing

Aircraft emit not only CO_2 but also water vapour, nitrogen oxide, soot and sulphate. Water vapour is a primary greenhouse gas, and nitrogen oxide reacts with sunlight to form another important greenhouse gas, ozone.

Because sunlight is more intense at high altitude, the impact of air travel is calculated to be 2.7 times greater than that of its CO_2 emissions alone, an amplification known as "radiative forcing". By this measure, aviation actually accounts for around 10% of man-made greenhouse-gas emissions. Flights in the Tropics, where the sun shines most brightly, are believed to be particularly damaging – a worrying finding, given the high proportion of fast-developing economies, such as India, in this part of the world.

CARBON OFFSETTING

Carbon offsetting enables air passengers to pay toward carbon-reducing projects to "offset" the emissions caused by their flight. However, many offsetting services are better at salving travellers' consciences than compensating for aviation emissions.

▶ FOCUS: Aviation and the economy

Proponents of aviation argue that a healthy air industry makes for a healthy economy. Not only are strong air links essential to a country's ability to trade internationally and attract tourists, but airports themselves are important economic entities: a major hub such as Amsterdam's Schiphol or New York's JFK can support as many as 500,000 jobs.

But the link between aviation and prosperity is not always clear. To stimulate growth, many countries effectively subsidize their air industry through low taxes on aircraft fuel and ticket sales. For example, in the UK in 2004, this amounted to £10.4 billion – almost as much as the £11.4 billion the country's air industry claimed to contribute to the economy.

Sources: NEF/World Development Movement

A greener future?

Airlines and aircraft manufacturers are working overtime to make flying more efficient, but are still some way from a lasting solution. The improvements they make are constantly being overtaken by increasing demand.

The technofix Due to enter service in 2010, Boeing's 787 Dreamliner will use 20% less fuel than existing aircraft of a similar size. In 2008, Virgin Atlantic proved it was possible to run a 747 on biofuel and aviation biofuels made from algae are on the (distant) horizon.

Lighten up Reducing drag with "winglets", removing surplus seats and stopping the common practice of "tankering" – overfilling the tank at airports where fuel is particularly cheap – can improve efficiency and help reduce emissions.

Smarter scheduling By taking more account of peaks and troughs in demand, airlines have been able to cut underused flights from their schedules. Fewer, fuller flights mean increased efficiency.

Rethink the need for air travel Advances in communications technology – from free internet telephone services to high-definition video conferencing – mean long-distance business travel can often be avoided. And vacationers can play their part by switching to road or rail and making the journey to their destination an enjoyable part of the experience.

FLIGHTS OF FANCY

Air spray
It has been suggested that planes could actually help reduce the effects of global warming. The idea is that aircraft would release aerosols into the stratosphere, in order to reduce levels of the greenhouse gas ozone. Even if this were possible, the side effects could be catastrophic (for example, exacerbating the problem of acidification of soils and seas) and unpredictable.
Source: Oliver Tickell, *Kyoto2* (2008)

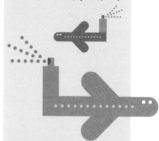

Steward-less
In response to the fierce competition for prime slots at the world's premier airports, aviation regulators have come up with a "use it or lose it" rule. On occasion, this policy forces airlines to run "ghost flights" with no passengers on board on less desirable slots – simply to protect their more prized slots.

ECO-FUELS

The modern internal combustion engine, run on petroleum or diesel, is more efficient than it has ever been. However, it is still a highly **wasteful** way of converting **chemical** energy to **kinetic** energy. Alternative fuels are on the horizon, but each has **important limitations** that will need to be resolved before oil disappears in the rear-view mirror of automotive history.

⬇⬆ Green light for green fuels

- However much car manufacturers reduce emissions from their internal combustion engines, the technology is **inherently inefficient** – only around 15% of the energy contained in the fuel is converted into **motion**, the rest is lost as **heat**
- **Cleaner**, more **efficient**, more **sustainable** fuels are coming on stream – as car sales **plummet**, eco-fuels offer **hope** for the future of the automotive industry

⬇⬆ Red alert – are eco-fuels ready?

- Some alternative technologies may offer hope, but the most promising are **many years away** from the mass market, while others, such as biofuels, seem to present **intractable problems**
- In the meantime, we should continue to improve the efficiency of the **internal combustion engine**

Biofuels

Biofuels are fuels that can be grown rather than mined. The organic matter is distilled and burned to drive an engine.

"First generation" biofuels come in two main types. Bioethanol is made from crops such as sugar cane and maize, often blended with petrol. Biodiesel is generally made from soya, palm-oil and rapeseed, often blended with diesel.

Many diesel engines will take biodiesel without modification. Otherwise, it's a simple process to convert engines.

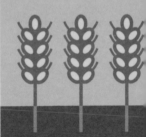

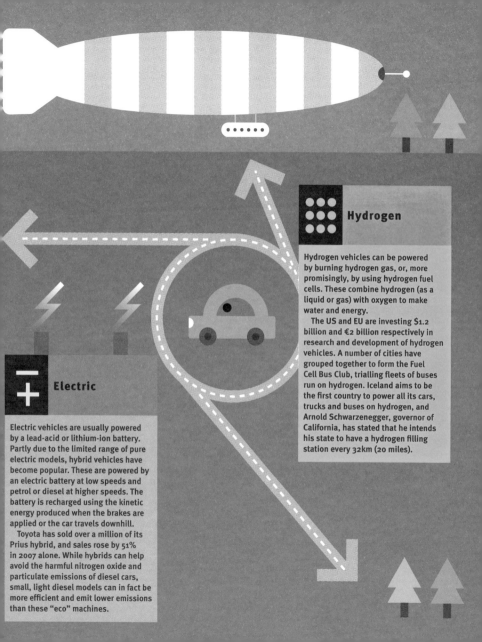

Hydrogen

Hydrogen vehicles can be powered by burning hydrogen gas, or, more promisingly, by using hydrogen fuel cells. These combine hydrogen (as a liquid or gas) with oxygen to make water and energy.

The US and EU are investing $1.2 billion and €2 billion respectively in research and development of hydrogen vehicles. A number of cities have grouped together to form the Fuel Cell Bus Club, trialling fleets of buses run on hydrogen. Iceland aims to be the first country to power all its cars, trucks and buses on hydrogen, and Arnold Schwarzenegger, governor of California, has stated that he intends his state to have a hydrogen filling station every 32km (20 miles).

Electric

Electric vehicles are usually powered by a lead-acid or lithium-ion battery. Partly due to the limited range of pure electric models, hybrid vehicles have become popular. These are powered by an electric battery at low speeds and petrol or diesel at higher speeds. The battery is recharged using the kinetic energy produced when the brakes are applied or the car travels downhill.

Toyota has sold over a million of its Prius hybrid, and sales rose by 51% in 2007 alone. While hybrids can help avoid the harmful nitrogen oxide and particulate emissions of diesel cars, small, light diesel models can in fact be more efficient and emit lower emissions than these "eco" machines.

Electric

Electric cars – climate-friendly and ready for the road?

- If the electricity they use comes from a renewable source, electric vehicles produce almost no carbon.
- Power stations tend to have excess generating capacity at night, which could be used for charging vehicles.
- Electric motors generally have around 90% fewer parts than conventional ones and are less likely to break down.
- They're relatively cheap to run once the infrastructure is in place, although electricity is lost in grid inefficiencies.

Some potential short circuits

- Batteries contain harmful substances such as cadmium.
- Due to the challenges of storing power, few fully electric cars have a range of more than 150km (90 miles). A full charge can take up to five hours.

HAILING HYBRIDS

In 2007, New York launched a five-year initiative to convert the city's 13,000 yellow cabs to more efficient hybrid vehicles. The plan, which requires all new vehicles entering the fleet to average at least 25mpg [US] (9.4l/100km) in urban driving, should avoid 200,000 tonnes of CO_2 emissions a year (or 10,000 New Yorkers' carbon footprints).

Ironically, the project is in danger of running behind schedule – not because of a lack of interest, but because hybrid manufacturers are struggling to keep up with high consumer demand.

Source: PlaNYC (2007)

Hydrogen

Healthy hydrogen – the low-carbon fuel of the future?

- Hydrogen is the most abundant element in the universe.
- A vehicle fuelled by hydrogen emits no greenhouse gas or other pollutants, just water.

Or expensive, energy-intensive and inefficient?

- Hydrogen is a carrier, not a source, of energy, because to make it into a fuel requires a huge energy input.

GOING THE DISTANCE

Every year since 1985 Shell has held an event called the Eco-Marathon, in which teams compete to design a vehicle that will travel the longest distance on the least fuel. The all-time record was set in 2005 when a Swiss-built vehicle powered by hydrogen fuel cell travelled 3,836km (2,384 miles) – the equivalent of driving from Paris to Moscow – on a single fuel cell.

Source: Shell Eco-Marathon

- Even when compressed or liquefied, a hydrogen fuel load is heavy, which reduces vehicle efficiency.
- Fuel cells are currently very expensive to produce and require rare materials such as platinum.
- Currently, fuel cells function poorly in cold weather.

IF AMERICA TURNED TO HYDROGEN ...

If the American transportation fleet were to use hydrogen instead of oil, every day the country would need 200,000 tonnes of hydrogen – enough to fill about 13,000 Hindenburg airships. If the electricity needed to produce this hydrogen were generated using wind power, an area the size of New York State would have to be covered with wind turbines.

Source: Paul Grant, "Hydrogen Lifts Off – With a Heavy Load", *Nature* (2003)

Biofuels

The great hope – renewable and climate-friendly?

- Biofuels emit less greenhouse gas than fossil fuels and have the potential to be "carbon neutral" and renewable.

Or land-hungry and greenhouse-gas heavy?

- Biofuels have contributed to recent food shortages by taking crops that could have been used as food.
- Converting peatland and rainforest to biofuel plantations releases methane and reduces the amount of CO_2 the land soaks up, meaning that biofuels can have a greater climate-change impact than fossil fuels.
- Monoculture biofuel plantations support little wildlife and can rapidly deplete soil fertility.

NEW GENERATION ON TRACK

The search is on for more suitable biofuels:
- Used as biodiesel for trains in India, jatropha is a hardy bush that can grow quickly on poor land. The Indian State Railway company has planted 7.5 million jatropha trees along its tracks.
- Marine algae is another possibility. It is reportedly capable of providing 15 times more biofuel for a given area than other crops such as rapeseed.

LIVING WITH CARS

It's projected that there'll be 3 billion motor vehicles on the world's roads by 2050 – that's more than three times the number currently in circulation. The **personal freedom** our cars can provide comes with **significant costs** – from congestion and pollution to resource depletion and ill health. If we **drive thoughtfully, sparingly and communally**, can we avoid the worst effects of climate change, or do we need to **break our car habit** to avoid total gridlock and climate chaos?

⇅ Carmageddon: the fast lane to chaos

- The environmental, social and economic **costs of car use** are overwhelming – from unsustainable **resource use** and **pollution** to growing **health problems** and the loss of productive time while stuck in **traffic jams**
- It's not only the vehicles themselves – **road-building** is resource-intensive and destroys ecosystems

- In 2005, cars were responsible for **14–17%** of global man-made CO_2 emissions – if car ownership continues to grow at current rates without cars getting any greener, by **2050** they'll raise global emissions so much they could single-handedly take temperatures 3°C (5.4°F) above pre-industrial levels, spelling ecological disaster

LAND OF THE RISING CLOUD

In 2002, American drivers produced more greenhouse gas emissions than the entire Japanese economy. However, the green shoots of improved auto-habits may be showing. Whereas in 2000 Americans bought more gas-guzzling SUVs than standard cars for the first time, in 2008 cars regained the top spot.

Sources: Environmental Defense Fund (2002)/Green Car Congress (2008)

⬆ Car sense: a more balanced way to use our cars

- Maintaining the current growth in car use is unsustainable, so we'll need to use cars more **cleverly** if more people are to enjoy **personal mobility** with lower environmental, social and economic costs
- Car engines are at their most **thirsty** when they are **cold**, so leaving the car at home for **short journeys** will have a disproportionately beneficial effect

- Low-cost alternatives to car ownership, such as **car sharing** and **car clubs**, give the convenience of car travel when it is most needed, but remove the temptation to use a car for every journey
- Adjusting driving style to maximize fuel economy (so-called **eco-driving**) is another way to use our cars responsibly – but better not to use them at all

CONGESTION CHARGING

A number of cities, including London, Singapore and Stockholm, have started to charge drivers to enter a designated zone at peak periods. Traffic entering London's original charging zone remains 21% lower than pre-charge levels, with 70,000 fewer vehicles a day. CO_2 emissions have been cut by 16% and traffic delays by 20%.

In Amsterdam and Copenhagen almost as many people commute by bike as by car, while 25% use public transport. This transport mix has been achieved through a combination of street design that encourages walking and cycling, taxes that incentivize public transport, car-sharing programmes and a change in public attitudes.

▶ FOCUS: Does extra road capacity reduce congestion?

You might think that building extra road capacity for a busy route would reduce congestion. Studies have shown that journeys do indeed get quicker at first, but that soon the new, improved route becomes a victim of its own success, as it encourages people to develop the habit of commuting by car more often or from further away.

Source: Newman and Kenworthy, *Sustainability and Cities* (1999)

In 1900 the average traffic speed in central London was 11mph (18kph). A century later, despite the rise of the car, average traffic speed had not changed.

The true cost of driving

KILLER SUVs

As well as using significantly more fuel and emitting up to 50% more greenhouse gases, SUVs are 27 times more likely than smaller cars to inflict fatal injury on people they hit.
Source: US Insurance Institute for Highway Safety

Prices at the showroom and fuel pump represent only a small proportion of the true cost of driving, for drivers and non-drivers alike. Every taxpayer contributes to the cost of building and maintaining road-transport infrastructure, and the healthcare of the victims of road accidents, vehicle-generated air pollution and sedentary lifestyles is another heavy liability.

One British study calculated that if all these hidden costs of driving were passed back to motorists as fuel tax, fuel would be six times more expensive. As it is, the whole population is effectively subsidizing road travel.

▶ FOCUS: The global rise of the car

Although perhaps not as expensive as it should be (see above), owning and running a car still requires a significant outlay, and currently only one in ten people worldwide drives. But car ownership is on the rise. One thousand more vehicles are hitting Beijing's streets every day, and by 2030 it's likely that China's fleet will have overtaken America's (which will also have grown by 60%), and by 2050 there will be as many cars in China as there currently are in the whole world. India will be catching up fast, with around 350 million cars on its roads by 2050 – that's 45 times current numbers.

This huge growth is fuelled by a combination of increasing wealth and falling prices. Cars like the US$2,500 Tata Nana, launched in India in 2008, are designed to do for the developing world what the Model T Ford did for America 90 years ago – make it normal for ordinary people to own a car.

🔢 Nine out of ten Americans and six out of ten Europeans and Japanese of driving age own a car, while the Chinese share three cars between every 100 people.
Source: *The Economist*

Do you need your own car?

If you have your own car, it can be hard to resist the temptation of jumping into it for a short journey that could easily be made by foot, bus or bicycle. More and more people are finding different, greener ways of enjoying the benefits of driving without owning a car.

Car clubs Pay-by-the-mile car-hire programmes, often known as car clubs, enable members to pick up a car at short notice from a nearby pickup/dropoff point. As well as saving themselves the overheads of car ownership, insurance, road tax and maintenance, members may also benefit from free parking bays and preferential driving lanes.

Car sharing In the UK, if every solo car commuter shared a lift to work at least once a week, traffic volumes would be cut by 12–15%. To facilitate communal travel, car-sharing organizations act as commuter "dating agencies", matching up subscribers who have compatible journey patterns.

ECO-DRIVING

Altering driving style and car-maintenance habits to improve fuel economy can have a significant impact. For example, driving at 50mph is 30% more efficient than driving at 70mph, and correctly inflated tyres can boost fuel economy by up to 10%.

Some countries, such as the Netherlands, Austria, Germany and Spain, run national eco-driving training programmes, which have been shown to be a cost-effective way of reducing CO_2 emissions. The fuel economy of participants in eco-driving courses improves by an average of around 10% after the training.

However, drivers can slip back into bad habits over time if they don't receive follow-up training or support. One effective way to provide continuous feedback on eco-driving performance is by installing dashboard instrumentation that displays fuel-economy readings.

Source: International Transport Forum

NEW FOR OLD

In a bid to kickstart its country's flagging motor industry, and also to take gas-guzzling old cars off the road, in 2008 the German government introduced a cash incentive of €2,500 for owners of vehicles more than nine years old to scrap their old car and replace it with a new, efficient model. This measure has been partly successful – sales have risen dramatically. However, environmentalists are less convinced of the policy's eco-credentials. They argue that there are many more cost-effective ways of achieving CO_2 reductions, and that, taking into account the energy embodied in the manufacture of the new vehicles, this subsidy may actually lead to an overall *increase* in emissions.

LIVING WITHOUT CARS

Despite being fully aware of the environmental costs, many of us jump into a car for almost every journey we make. Drivers often claim that alternatives are unreliable, slow, expensive or non-existent. However, after some years of underinvestment in many areas, public transport is mounting a strong recovery. Smarter, more convenient, lower-carbon alternatives to driving (and flying) are coming on stream all the time. Will they be good enough to tempt us out of our cars?

⬇️⬆️ Ditching the car habit is liberating

- Walking, cycling and public transport are greener and healthier than car travel
- Our reliance on cars is not a necessity, but a deeply rooted habit fed by government policy and investment priorities
- More roads simply means more gridlock with private cars – mass transit uses limited space much more efficiently

- Studies show that we tend to underestimate journey times by car and overestimate journey times by public transport
- Public transport is enjoying a renaissance, with policy-makers mindful of its environmental, social and economic benefits

ALL ABOARD THE WALKING BUS

Parents drive their children to school to protect them from busy roads, but if they didn't drive the roads would be much quieter and safer (in the UK, for example, the "school run" accounts for 20% of peak-time traffic). The "walking bus" is a greener solution – it involves children walking to school in a supervised group along a set route.

⬆In the modern world, the car is king

- In some locations public transport can be **infrequent, unreliable, expensive and slow**
- Public transport is often most effective in **densely populated** urban settings, and can be less practical in **sparser areas** or where amenities are located out of town
- In some areas car-focused street design makes walking and cycling **impractical and dangerous**

- Having your own car gives you **freedom and flexibility** that public transport can't rival

> ❗ In the UK between 1975 and 2003, the cost of bus and coach travel rose in real terms by 66% and train travel became 70% more expensive, whereas the cost of running a car fell by 11%.
> Sources: Office of National Statistics/Dept of Transport

▶ FOCUS: The daily grind

The total distance travelled in the UK more than trebled between 1952 and 1996, from 218 to 719 billion passenger km (135 to 447 billion miles). Much of this resulted from people choosing to live further away from where they work. Life-long London commuters will spend an average of a year and a half of their lives just travelling to and from work. Around 3.4 million Americans have a daily commute of 90 minutes or more, primarily by car – the average American adult spends 72 minutes a day in a car.
Source: DTI/Worldwatch Institute

> ❗ The average American household spends 17% of its annual income on owning and running motor vehicles.
> Source: Bureau of Labor Statistics

HIGH-OCCUPANCY VEHICLE LANES

Priority lanes for high-occupancy vehicles are designed to incentivize car-sharing. In the US the mutually beneficial practice of "slugging" has grown up around High-Occupancy Vehicle (HOV) lanes. This involves solo drivers calling out their destination to a queue of commuters and picking up those who want to go to the same place.

However, some studies have concluded that HOV lanes lead to greater levels of congestion and pollution in the other lanes, while failing to increase car-sharing; others have identified a heightened risk of accidents caused by fast-moving traffic running alongside slow-moving traffic.

Beyond the car

Trains

In the mid-twentieth century, the dramatic rise of air and road travel hit the passenger rail industry hard. In the West, countless small towns and villages lost their rail stations. For example, between 1963 and 1973 the UK rail network shed around 25% of its route miles and 50% of its stations. Although some of these closures have since been reversed, rails come a poor second to roads in terms of coverage – the world's roads cover a distance 60 times longer than its railroads.

However, rail has a vital role as a fast, comfortable, convenient – and low-carbon – way of travelling between large cities. Unlike airports, train stations tend to be located in the heart of cities, and rail travellers

CASE STUDY: CURITIBA

The Brazilian city of Curitiba has a model modern bus network. All major roads have express bus lanes, and tree-shaded transfer terminals have become places of social activity. A flat rate for tickets applies across the network.

Since 1974, when Curitiba's mass-transit campaign began, the city's population has tripled, but its car traffic has fallen by 30%. Three-quarters of all commuters get to work by bus, and Curitiba has the cleanest air of any Brazilian city. The system is now self-financing and experts estimate it cost just 1% of an equivalent underground system.

▶ FOCUS: How does sea travel compare?

Passenger ferries emit less CO_2 per person per mile than almost any other form of fuel-driven transport. But, of course, they make up only a small percentage of the ships on the sea. Over 90% of the world's traded goods travel by cargo ship. These vessels emit more sulphur dioxide (the cause of acid rain and haze) than all the world's cars and lorries, but CO_2 emissions from international shipping account for "only" 2.7% of the global man-made total – significantly less than CO_2 emissions from cars and planes.

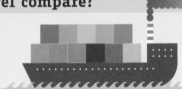

An ancient technique with a new spin has the potential to make modern cargo ships more environmentally benign: new ships can operate computer-controlled sails as big as football pitches, reducing their fuel use by up to 35%. The most advanced sails can also generate electricity from thousands of tiny solar panels which track the sun.

Source: SkySail (2008)

can turn up five minutes before departure. And, unlike cars, the fastest trains in the world, such as the Japanese Maglev and French TGV, can reach speeds of more than 560kph (350mph).

Buses

The energy-efficiency of any form of transport has a lot to do with the number of passengers carried – a full bus is twice as efficient as a half-full one. In some countries the traditional bus that travels along a set route at set times scores poorly by this measure. In the US in 2006, the average number of passengers per bus was just 8.8.

However, countries such as the Netherlands and Germany are pioneering flexible approaches to bus travel. For example, the "taxibus" is pre-booked like a taxi and provides a door-to-door service, but it also picks up other passengers, using computer software to group customers with compatible journey requirements.

Bikes

Cycling is the most environmentally benign form of mechanized transport. Although cyclists need to consume more calories than drivers to fuel their efforts, greenhouse gas emissions from cycling are still minimal. What's more, studies show that people who commute by bicycle tend to be happier with their journey than those who drive.

Improved facilities such as dedicated cycle lanes are tempting more and more people onto their bikes; in the US about 12% more cyclists join the roads every year.

TWO WHEELS GOOD

Pioneering Paris
The hugely successful "Vélib'" bike hire programme in Paris has over 20,000 bicycles parked in more than 1,500 bays across the city. For a small fee, subscribers can pick up a bike from one bay and drop it off in another. As a result of Vélib', cycling in Paris went up 68% between 2006 and 2007. Similar programmes operate in other cities, including Barcelona, Copenhagen and Vienna, and are planned in London, Rome, Moscow and Beijing.
Source: Vélib' (2007)

Sort out your priorities
Cyclists have priority on Dutch roads, perhaps explaining why, per km, there are 30 times fewer cycling injuries in the Netherlands than in the US, and eight times fewer deaths.
Source: Worldwatch Institute

CHAPTER 5

EAT, DRINK, SHOP

It's a chore for some and a pleasure for others, but we all need to shop. A growing number of ethical consumers are taking advantage of their purchasing power to lobby retailers to become greener. In a world where more people are overweight than undernourished, a potato can travel further than Christopher Columbus from field to checkout, and a t-shirt often costs less than a loaf of bread, can the way we choose to feed and clothe ourselves really make a difference?

In this chapter we'll chew over the following questions:
- Can organic farming feed the world?
- Why is trade unfair, and is Fairtrade any fairer?
- What's the beef about meat and fish?
- Where should the ethical consumer buy groceries?
- Is eco-chic a passing fad or fashion's future?

ORGANIC FARMING

The world organic market has been growing by 20% a year since the early 1990s, as more and more consumers seek an alternative to food produced intensively with the aid of petrochemical-based pesticides and fertilizers and plentiful antibiotics. But debate still rages between supporters of organic farming and those who believe that only intensive agriculture can satisfy the appetite of a growing population.

⬇⬆ The planet needs to go organic

- By promoting biodiversity and healthy soil, organic farming works with nature rather than fighting against it
- Food produced organically is largely free from pesticide residues and can be more nutritious than conventionally produced food

- Intensive farming is unsustainable – repeated applications of energy-intensive fertilizer and pesticide lead to soil degradation and in the worst cases can even result in desertification
- Genuine organic farming respects the welfare of livestock and farmworkers

ORGANIC FARMS HAVE UP TO:

85% more plant species

33% more bats

17% more spiders

5% more birds

Source: BTO et al (2005)

WHY IS BIODIVERSITY IMPORTANT?

Complex ecosystems with a wide variety of plants and animals tend to be more stable – more resistant to disease, infestation and drought, and less prone to soil erosion. Organic agriculture seeks to foster this diversity by:
- using compost to feed the numerous earthworms and micro-organisms to be found in healthy, fertile soil
- providing habitats such as hedgerows and "companion plants" next to crops to attract natural predators of crop pests (thereby minimizing the need for pesticides)
- growing a wide range of crop varieties

⬆ Only conventional farming can feed the world

- Without improvements in organic yields, we'll struggle to provide for current global food consumption 100% organically
- Organic standards and ideals are at risk of being watered down as large **corporations** get more involved

- Demand in the major markets for organic food often exceeds local supply, leading to a high proportion of **imports**
- Research into the supposed health benefits of organic food remains **inconclusive**

THE "GREEN REVOLUTION"

Using intensive methods, global cereal production tripled between 1950 and 2000, but only 10% more land was required.

1950   2000

In the mid-twentieth century, faced with the problem of how to feed a rapidly growing population, agronomists developed the principles of intensive farming: a focus on the highest-yielding crop varieties, economies of scale, and greater use of technology and chemicals.

FOCUS: The killing fields

An unforeseen consequence of increased fertilizer use is the growth of marine "dead zones" – areas of coastal waters where oxygen levels are too low to support aquatic life. This phenomenon stems from an excess of nutrients (known as eutrophication), which can be caused by fertilizer pollution. Is the gap in yield between intensive and organic farming great enough to justify this degree of ecological damage?

🔲 In the developed world average yields from organic land are less than 10% lower than yields from intensively farmed land and they are 200–300% higher in the developing world.

Source: University of Michigan (2007)

1.5 million tonnes of fertilizer run off Midwest cornfields into the Mississippi each year. They flow into the Gulf of Mexico where they help to create a 22,000 km^2 (8,500 miles^2) "dead zone" ...

... this is an area the same size as the state of New Jersey.

Is organic food healthier?

A four-year comparative study which published its results in 2007 found that organic food contained on average 40% more antioxidants – vital in protecting against cancer and heart disease – than conventionally farmed equivalents, as well as higher levels of minerals such as iron and zinc. However, for every piece of research that finds in favour of organic food, there may be another suggesting there is little or no nutritional difference between organic and non-organic food.

It *is* generally agreed that organic food contains lower levels of pesticide residue. The official advice is that the amount of residue in or on non-organic food is too small to cause any harm, but the organic lobby argues that the effects of each of these compounds have been tested only individually and that we have no way of knowing how they affect our bodies when ingested as a "cocktail".

PERCEPTION IS EVERYTHING

Reasons given by consumers for purchasing organic food:

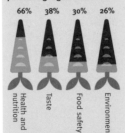

66% 38% 30% 26%

Health and nutrition | Taste | Food safety | Environment

The ecological case in favour of organic food is better established than the nutritional one, yet concern for the environment lags a long way behind perceived health benefits in consumer surveys.

Source: The Hartman Group (2006)

▶ FOCUS: The white stuff

A number of comparative studies suggest that organic milk contains higher levels of certain vitamins and minerals than its conventionally farmed equivalent (see right, for example). One piece of research observed a rise in omega 3 content in organic milk during the months when the cows were put out to pasture and also hypothesized that certain traditional breeds, such as Jerseys, largely rejected in conventional farming for their relatively low yield, may produce more nutritious milk.

Organic milk contains

50% more Vitamin E

64% more omega 3

75% more beta-carotene

Source: Danish Institute of Agricultural Nutrition

Organic in a changing climate

Which is responsible for more greenhouse-gas emissions per hectare: organic farming or intensive farming?

On the one hand, low-energy techniques (particularly the avoidance of fertilizer made using fossil fuels, which accounts for nearly 40% of the energy used in non-organic agriculture) mean that organic farms tend to emit less CO_2 per hectare than conventional counterparts. For example, organic milk requires only a third of the primary energy input of non-organic milk. Another factor in favour of organic farming is that organically farmed soils capture and store up to 30% more CO_2 than soils on conventional farms.

On the other hand, some agronomists argue that organic farming is responsible for higher levels of certain other greenhouse gases than conventional farming – mainly because organic yields are lower. Looking at organic milk production again, lower yield means that more cows are required to produce the same amount of milk. More cows result in more methane, as well as more resources required to produce more feed. A study by Cranfield University in the UK claims that a litre of organic milk actually has 20% *more* impact on the climate than a litre of non-organic milk.

However, it's almost impossible to come up with a truly holistic answer that takes into account hard-to-quantify factors such as biodiversity and soil health – areas where organic farming scores particularly highly.

PEST CONTROL

Pests must be kept at bay even on organic farms, and if crop rotation, companion planting and other biological controls have failed, a very restricted list of pesticides can be used on organic crops – but only as a last resort to solve particularly stubborn infestations.

Under the UK's Soil Association standard, only copper, rotenone, soft soap, sulphur and Bt (Bacillus thuringiensis) can be used, and then only for very specific purposes. Rotenone was allowed by the USDA National Organic Program (American standard)

Organic farming uses 97% less pesticide per acre than conventional farming

until 2005, when it was removed from the list. These substances can harm beneficial or innocuous organisms such as bees and butterflies and their effects on human health haven't been rigorously tested, so it's important that their use remains strictly limited.

FAIRTRADE

The "Fairtrade" network achieved prominence during the "coffee crisis" of the 1990s and now supports millions of producers in the developing world. Is the premium we pay for Fairtrade goods **well spent**? Or does Fairtrade distract from the bigger task of making **all trade** fair?

⬇⬆ The world needs Fairtrade

- Fairtrade subsidies are designed to be a **necessary corrective** to the much greater support given to producers in the EU and US
- As well as subsidizing producers, Fairtrade often gives them the **capital, advice and support** they need to diversify into products for which there is ample demand
- Fairtrade and similar programmes bring attention to trade issues and **raise expectations** of how all producers should be treated

⬇⬆ Fairtrade isn't fair

- Some critics argue Fairtrade **floods the market** and reduces overall prices by paying producers for what they make even when nobody's buying it, leading to small-scale, unambitious production and poor-quality products
- There have been claims that producers still don't get as large a profit share as they should
- Fairtrade is accused of being a **distraction** from the bigger job of changing international trade rules

A fair exchange …

For products to carry the "Fairtrade" label ("Fair Trade Certified" in the US), producers must be in a democratic co-operative group, sharing responsibility and profits. They must also conform to environmental standards and treat workers fairly, employing no child labour.

… is no robbery

In return for their commitment to uphold Fairtrade standards, producers are guaranteed a decent fixed price for their certified goods whatever happens on the world market, receive a premium to invest in the community and are assured a long-term, direct relationship with buyers.

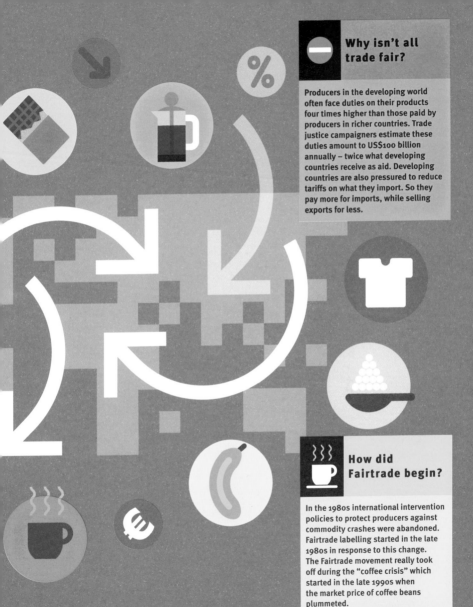

Why isn't all trade fair?

Producers in the developing world often face duties on their products four times higher than those paid by producers in richer countries. Trade justice campaigners estimate these duties amount to US$100 billion annually – twice what developing countries receive as aid. Developing countries are also pressured to reduce tariffs on what they import. So they pay more for imports, while selling exports for less.

How did Fairtrade begin?

In the 1980s international intervention policies to protect producers against commodity crashes were abandoned. Fairtrade labelling started in the late 1980s in response to this change. The Fairtrade movement really took off during the "coffee crisis" which started in the late 1990s when the market price of coffee beans plummeted.

Spending the premium

Some consumers are reluctant to pay extra for Fairtrade, suspecting that very little of the premium finds its way to the producers. It is true that the retail margin for Fairtrade products is generally higher – retailers argue that this reflects the higher costs involved in giving shelf space to Fairtrade products, because they tend to sell in lower volumes and at a slower rate than "own-brand" equivalents.

But Fairtrade producers do receive a premium, and there are many examples of this money being invested in valuable community projects, such as health programmes or improvements to local schools. In other cases single-crop producers use the premium to protect themselves against future commodity crashes by diversifying into other crops.

THE 7.5 MILLION BENEFITS

It's estimated that 7.5 million people directly benefit from Fairtrade, with over 600 certified producer organizations in 58 countries, representing 1.5 million farmers and workers. Worldwide sales total around US$2.9 billion. Sales in the UK – where one in every four bananas sold is Fairtrade – doubled between 2007 and 2008.

Source: Fairtrade Labelling Organization

► FOCUS: How eco is Fairtrade?

Fairtrade producers have to commit to respecting "a balance between environmental protection and business results". They must minimize their use of synthetic fertilizers, reduce, reuse, recycle or compost waste, maintain soil health and use water and energy efficiently. Generally, it is not in farmers' interests to farm unsustainably, as they have long-term relationships with Fairtrade buyers and only limited land. However, Fairtrade should not be confused with organic production, which has more stringent environmental standards.

Is Fairtrade counterproductive?

According to a 2008 report published by the influential economic think-tank the Adam Smith Institute, the answer is most definitely "yes". The report, entitled "Unfair Trade", argues that a guaranteed minimum price offers little incentive to Fairtrade producers to improve their efficiency, for example by mechanizing their farms. Guaranteed minimum prices can also lead to oversupply, which distorts the market and can lead to the kind of crash that Fairtrade was supposed to protect against. According to the report, although Fairtrade shelters its own producers from price fluctuations, the much larger group of non-Fairtrade growers are often left out in the cold. Rather than trying to correct one distortion of the free market (tariff barriers) by introducing another (guaranteed prices), critics of Fairtrade argue we should campaign for the removal of the tariffs.

Beyond Fairtrade

Other programmes supporting growers in developing countries work differently. For example, Starbucks' own system, CAFE (Coffee and Farmer Equity), doesn't offer producers a fixed price, but does give them preferential supplier status if they meet certain independently verified social, environmental and quality standards.

As well as, or instead of, buying Fairtrade (or similar) products, consumers can also directly fund small-scale entrepreneurs in developing countries via the micro-lending website kiva.org.

BEAN COUNTING

Of the US$55 billion-worth of world coffee sold every year, little more than $5 billion finds its way back to the growers. This is how the final price of a typical jar of coffee is distributed:

Retailers: 25%

Shippers and roasters: 55%

Exporters: 10%
Growers: 10%

Source: Oxfam

Adding value
As the figures above suggest, processing and selling food and drink can be much more profitable than growing it. While they receive a small premium, Fairtrade producers don't share in this "added value" – raw materials are still exported to developed countries.

The Costa Rican company Café Britt is an unusual example of a coffee roaster and retailer based in the country of production. Its suppliers are allowed to benefit from the more profitable parts of the process.

ANIMAL OR VEGETABLE?

In the West we eat many times more meat, fish and dairy than the global average. Owing to the major environmental impacts of fish- and livestock farming, it has been calculated that the average American could do more to reduce his or her carbon footprint by **becoming a vegetarian** than by switching to a hybrid car. Can we **eat away at meat** without it eating away at us, or are greens the only **green option**?

⬇⬆ Go vegetarian to save the planet

- Livestock farming is much more **land-, water- and energy-intensive** than arable farming
- Energy is lost at each level of the **food chain,** so it's much more efficient to eat plants rather than the animals that feed on them
- It's difficult to ensure that the meat you eat is produced according to high animal **welfare standards** – and even the most humanely reared animals are still killed for our gastronomic pleasure

- Livestock produce large amounts of **methane,** a potent greenhouse gas
- **Overfishing** is bringing many fisheries to the point of collapse
- **Farmed fish** need vast amounts of food, are vulnerable to disease and can harm local fish populations
- **Vegetarians** generally have lower rates of obesity, coronary heart disease, high blood pressure, large bowel disorders and cancers than their carnivorous counterparts

PLENTY MORE FISH OUT OF THE SEA

By the time industrial fishing came of age in the early 1980s, every two years the world's fishermen were catching as many fish as caught in the whole of the 19th century. Demand for meat has followed a similar trajectory: worldwide, per capita meat consumption has risen by 500% since 1950.

↑ Eat meat and fish, but choose carefully

- In some cases, grazing livestock on pasture land can **increase biodiversity** and **optimize the productivity** of marginal farmland
- Consumer pressure is encouraging **more humane** farming, transportation and slaughter of animals
- Farming is key to the survival of some **rare breeds**
- Methane from manure can be harnessed for **biofuel** by means of anaerobic digestion (see page 35)
- Animal products are particularly rich sources of certain important **nutrients**

LOOK BEYOND THE LABEL

Ethical meat-eating requires a certain amount of effort (some would argue that it's impossible). To be confident that the meat you're buying has been humanely reared, shop at a local farmers' market where you can speak directly to the producer. Local, small-scale farms often give animals access to pasture land and transport them directly to local abbatoirs, which reduces stress, and are more likely to stock rare breeds, which promotes genetic diversity.

FOCUS: The "nutrition transition"

As inhabitants of developing countries become wealthier and more urbanized, they eat more meat and dairy products. The table below illustrates this so-called "nutrition transition" in developing countries (units: kg per person per year).

	1962	1970	1980	1990	2000	2003
Cereals	132	145	159	170	161	156
Roots and tubers	18	19	17	14	15	15
Starchy roots	70	73	63	53	61	61
Meat	10	11	14	19	27	29
Dairy	28	29	34	38	45	48

Source: UN Food and Agriculture Organization (2006)

Playing cod with fish stocks

Overfishing has left 11 of the world's 15 major fishing areas and 69% of major fish species in decline – stocks of large fish, such as halibut, marlin, swordfish, sharks and tuna, have been particularly badly hit.

It's not only fish that are affected. Driftnets scoop up marine life indiscriminately, often trapping birds, coral and dolphins (it's estimated that at least 8% of the global sea-fishing catch is "bycatch"). Falling fish stocks affect whole food chains, with knock-on effects for entire marine ecosystems.

Consumers can help by reducing their fish consumption and buying fish that carry the Marine Stewardship Council's "tick" logo, a reliable sustainability certification for sea-caught fish.

HEALTHY HEARTS

Studies of Japanese and Inuit fishing communities have found a link between high consumption of oil-rich fish and low incidence of coronary heart disease. However, in order to be able to share in this health benefit, we ought to ensure our intake remains moderate.

▶ FOCUS: Fish farms

Fish farms are booming in response to our insatiable demand for seafood. Farmed fish are often kept in crowded conditions, making them vulnerable to disease. They need vast quantities of food (see right), which must itself be taken from the sea. Waste is the other major problem with aquaculture – fish farms deposit huge amounts of phosphorous, antibiotics, uneaten food and dyes directly into aquatic ecosystems, damaging wild fish stocks.

If you want to reduce the ecological impact of your fish consumption, buy certified organic fish. They're still farmed, but in much less intensive conditions, which means they can thrive without chemical intervention.

Approximately three kilograms of forage fish go to produce one kilogram of farmed salmon; the ratio for cod is 5:1; and for tuna – the most beef-like of all – the feed-to-flesh ratio is 20:1. Tilapia and catfish have better ratios.

Should I give up meat?

Undeniably resource-intensive and often inhumane, livestock farming fares badly when viewed in environmental and ethical terms. But is it healthy to remove animal products from our diets?

In some ways it's much healthier not to eat animal products. Studies identify a correlation between high consumption of meat and dairy and high incidences of coronary heart disease and hypertension, as well as certain types of cancer, such as cancer of the colon. (It is claimed that there has never been a case of colon cancer in a total vegetarian.)

However, vegetarians' health can suffer if they don't eat a balanced diet containing plenty of protein-rich food such as legumes. Vegans must eat cereals fortified with Vitamin B12 or take supplements, as this nutrient only occurs naturally in animal products.

It's wrong to think that a vegetarian diet is inherently healthier than an omnivorous diet (or vice versa). The key is to eat a variety of foods and avoid processed products. For this reason, some people have taken a "third way" – treating themselves to occasional meals of ethically sourced meat.

🔋 Australia exports around 6 million live sheep to the Middle East every year. Having travelled up to 50 hours by road to get to the sea port, they journey up to three weeks by sea and then continue by road at the other end.

CARBON HOOF-PRINT

The world is currently inhabited by twice as many chickens as people, 1 billion pigs, nearly 1.5 billion cows and 1.8 billion sheep and goats. Farming animals accounts for 18% of global man-made greenhouse-gas emissions (from livestock flatulence, industrial feed production, and transporting animals and carcasses). That's more greenhouse gas than produced by all forms of transport. Livestock farming also requires a huge amount of water (see below), most of which goes into feed production.

Livestock's mineral-rich effluent provides a useful boost to soil fertility in mixed, closed-system farms. But effluent from intensive livestock farming gets into waterways, with dire consequences for ecosystems.

It takes an average of 100,000 litres (22,000 gallons) of water ...

... to produce 1kg (2.2lb) of beef

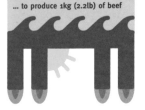

BUYING GROCERIES

We've become used to a mind-boggling variety of food sourced from all corners of the world and brought to a large air-conditioned building with plenty of parking. Recently this way of shopping has taken a greener turn; but do "eco" initiatives such as reusable carrier bags and biodegradable packaging make supermarket sweeps an ethical way of feeding ourselves? Or should we give our custom to local producers?

⬆⬇ Supermarkets: powerful forces for change

- Supermarkets provide a cornucopia of foods and groceries – many of which couldn't be produced locally
- By buying food abroad they support growers in the developing world
- Providing everything under one roof may reduce shopping miles
- Improved product labelling helps consumers make informed purchases

- Many supermarkets now compete for eco-conscious consumers by continuously improving their green and ethical standards
- With their huge purchasing power, supermarkets can instigate real change by insisting on higher ethical standards – and ideally supporting producers in implementing them

GREEN GROCERS?

Almost all the major world supermarket chains have made green pledges, from general aspirations to specific goals, such as Wal-mart's ambition to run its stores on renewable energy. One of the most wide-ranging projects is UK chain Marks and Spencer's "Plan A", launched in 2007. By 2012, the retailer aims to: become carbon neutral; send no waste to landfill; source more products sustainably; improve the lives of people in its supply chain; and help customers and staff live more healthily. In the first year CO_2 emissions from stores and offices were cut by 55,000 tonnes, three pilot "eco-stores" were opened, and all packaging was labelled according to its recyclability.

⬆ Local producers: a sure route to high standards

- A direct relationship with producers means the consumer can feel confident about where their food has come from
- **Solid and long-lived** relationships between producers and consumers provide more incentive for producers to implement positive change than the often **faceless and transitory** arrangements with supermarkets
- Local production reduces dependence on international markets and is essential for long-term **food security**, as well as providing a steady basis for the local economy

PUTTING DOWN ROOTS

Community Supported Agriculture (CSA) is a growing phenomenon in the US. There are similar schemes in other countries, including Canada, Japan and Germany. CSA involves consumers investing in a local, often organic, farm's harvest and receiving weekly shares of the produce in return for their commitment. Some programmes also give subscribers a say in the running of the farm; farmers may even welcome a spare pair of hands at busy times.

CSA provides a guaranteed up-front source of revenue to local farms, which can help them to stay afloat during lean times. However, the downside for the subscribers is that if the harvest is poor they may receive less produce each week, or a less varied selection than they had been expecting.

▶ FOCUS: Logistics statistics

In the UK alone, supermarket lorries travel 657 million km (408 million miles) a year – the equivalent of going to the moon and back 854 times. However, some supermarkets are finding ways to cut back on distribution mileage. For example, ASDA (Wal-mart's UK division) has created "local hubs", which enable small-scale suppliers to pool their resources and coordinate their deliveries to the supermarket. In 2007, ASDA claimed that this initiative had saved 7 million road miles.

Sources: Norman Baker MP, *How Green is Your Supermarket?* (2004)/ASDA (2007)

Seasonality and food miles

Supermarkets' complicated sourcing, processing and distribution webs mean that even food grown relatively locally may have zig-zagged hundreds of miles before it reaches the shelf. The biggest problem with food miles is their contribution to carbon emissions. It has been calculated that air-freighting a single basket of produce from various parts of the world can emit around the same amount of CO_2 as an average four-bedroom household does through cooking all its meals for eight months.

Locally produced food doesn't always have a lower ecological impact than produce from far-flung places. In response to demand for all-year-round fruit and vegetables, many farmers in colder climates are growing summer crops such as tomatoes in heated greenhouses in early spring or late autumn. This can use at least as much energy as transporting them from balmier climes.

As a rule of thumb, buying locally produced food that's in season has the lowest impact, and encourages better farming practices at home. What's more, fresh, home-grown seasonal produce tends to be more nutritious than food transported thousands of miles, and is less likely to require refrigeration or elaborate protective packaging; and more varieties are available – not just the ones that can withstand long trips. Buying direct from farm shops or farmers' markets gives a greater return to the producers, and cuts down on distribution mileage.

Beyond food miles

Food miles may not always be the best measure of food's environmental impact. Researchers in New Zealand contend that rearing lamb in Britain for the British market is less energy-efficient than importing the meat thousands of miles from New Zealand, mainly because New Zealand sheep farmers use more renewable energy and less fertilizer.

Source: Lincoln University, Christchurch (2006)

Where your money goes

One study found that only 26% of the cost of a shopping basket of food purchased at the average UK supermarket is accounted for by the food itself; the rest goes toward packaging, processing, transportation, store overheads and advertising – not to mention the supermarket mark-up, which can be as high as 45%.

Source: National Farmers' Union (2004)

Keeping up appearances

By insisting on "perfect" produce with a uniform look, supermarkets force suppliers to waste huge amounts of edible food. For every tonne of bananas shipped, around two tonnes of fruit are left behind. Some supermarkets are beginning to relax their appearance criteria. Alternatively, shoppers can buy direct from local growers to avoid contributing to this wastage.

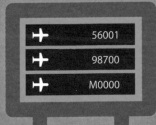

✈	56001
✈	98700
✈	M0000

The international livestock swap

Each year millions of live farm animals are transported between countries. As well as leading to suffering on the part of the animals and a huge carbon footprint, this exchange of livestock increases the risk of disease transmission. One of the main purposes of the trade – to increase genetic diversity – could be served by transporting semen or embryos instead.

APPROPRIATE APPAREL

Green credentials are now a major selling point in mainstream clothes retailing. But is this just a **passing fad**? Will sustainable labels end up as curiosities in tomorrow's rummage sales? Or will **eco-chic** show how clothes can really weather the seasons?

⇅ The new black

- Concerned consumerism continues to rise dramatically – an ever-increasing proportion of shoppers factor **ethical issues** into their purchasing decisions
- Internet shopping, **sharing, giving and swapping** sites allow clothes to be easily reused

- New sustainable materials such as hemp and organic cotton and wool are proving every bit as hardwearing, versatile and attractive as conventional materials
- Designers are rediscovering **natural dyes** for a mass market – and making their mark

⇅ So last year (and still here)

- Cotton – the world's most widely used natural fabric – is intensively farmed using **toxic pesticides** which can harm workers and enter the foodchain
- Unfortunately, the **sweatshop and child labour** are still a fact of fashion – although most big brands are now more transparent about their sourcing, consumers must be wary
- Despite the success of high-profile campaigns, **animals** are not necessarily treated fairly by fashion

WATCH OUT FOR FOOL'S GOLD

Gold mining leaves a shocking trail of destruction: the production of a single gold ring results in up to 20 tonnes of mine waste. Cyanide and mercury compounds are used to separate gold from its ore. As well as harming workers, they can pollute land and water, often affecting an area for long after the mine has closed.

Vintage or recycled gold jewelry is the lowest-impact option.

Getting it right for animal rights

Making sure animals get a better deal from your wardrobe doesn't require stripping off completely if you follow a few simple guidelines:

Buy a sheep-friendly jumper Some knitwear companies have an anti-mulesing policy. Mulesing is the practice (common in Australia) of cutting flesh from around the bottoms of sheep (and cashmere goats). If your favourite brand doesn't have such a policy, make your views known.

Look for alternatives to animal-sourced materials You can even get gorgeous handbags made from rubber tapped from Amazonian trees, which looks very much like leather.

Hire a handbag If only a leather or skin statement bag will do, borrow one from a bag library.

DON'T SHOP, SWAP!

"Swishing" is the name given to the growing trend of swapping clothes – an eco-friendly way of dressing on the cheap, turning used clothing into a valuable currency.

Swishing can be done either in person at dedicated parties, or online. Either way, participants enjoy the satisfaction of acquiring a free new wardrobe while passing on those clothes that were never quite the right fit or colour.

CUTE COTTON?

Cotton growing accounts for about 22.5% of global insecticide use, and 10% of pesticide use, although it covers only about 2.5% of the world's agricultural land. These chemicals can kill wildlife, contribute to climate change and contaminate drinking water. They're not great for people either – 20,000 die each year from pesticide poisoning, many in cotton production. Another 3 million or so suffer side effects including cancer, birth defects, respiratory problems, infertility and sterility. So buy organic!

Global retail sales of organic cotton were projected to grow to US$2.6 billion by the end of 2008, which is 0.19% of the 25.5 million tonnes of cotton produced each year.

FURTHER READING AND USEFUL WEBSITES

FURTHER READING

Michael Braungart and William McDonough, *Cradle to Cradle*. North Point Press, 2002

Lester R. Brown, *Plan B 3.0*. Earth Policy Institute, 2008

James Bruges, *The Big Earth Book*. Alistair Sawday Publishing, 2008

Rachel Carson, *Silent Spring*. Houghton Mifflin, 1962

Duncan Clark, *The Rough Guide to Ethical Living*. Rough Guides, 2006

Herman E. Daly and John B. Cobb Jr, *For the Common Good*. Beacon Press, 1989

Anne H. Ehrlich, *The Dominant Animal: Human Evolution and the Environment*. Island Press, 2008

Thomas L. Friedman, *Hot, Flat and Crowded: Why the World Needs a Green Revolution – and How We Can Renew Our Global Future*. Farrar, Straus & Giroux, 2008

James Garvey, *The Ethics of Climate Change*. Continuum, 2008

Al Gore, *An Inconvenient Truth*. Rodale, 2006

Paul Hawken, *The Ecology of Commerce: A Declaration of Sustainability*. Collins Business, 2005

Paul Hawken, Amory B. Lovins and L. Hunter Lovins, *Natural Capitalism: The Next Industrial Revolution*. Earthscan, 1999

Robert Henson, *The Rough Guide to Climate Change*. Rough Guides, 2006

Thomas Homer-Dixon, *The Upside of Down: Catastrophe, Creativity and the Renewal of Civilization*. Island Press, 2008

Ivan Illich, *Energy and Equity*. Marion Boyars, 1974

IPCC, *Fourth Assessment Report: Climate Change 2007*. Cambridge University Press, 2007

Nigel Lawson, *An Appeal to Reason: A Cool Look at Global Warming*. Duckworth Outlook, 2008

Bjorn Lomborg, *Cool It: The Skeptical Environmentalist's Guide to Global Warming*. Alfred A. Knopf, 2007

James Lovelock, *The Vanishing Face of Gaia*. Allen Lane, 2009

Bill McGuire, *Seven Years to Save the Planet*. Orion, 2008

Nick Middleton, *The Global Casino*. Hodder Arnold, 2003

George Monbiot, *Heat: How We Can Stop the Planet Burning*. Penguin/South End Press, 2007

Mark Niemeyer, *Water: The Essence of Life*. Duncan Baird Publishers, 2008

Jonathon Porritt, *Capitalism as if the World Matters*. Earthscan, 2006

Sonia Shah, *Crude: The Story of Oil*. Seven Stories Press, 2004

Andrew Simms and Joe Smith, *Do Good Lives Have to Cost the Earth?* Constable & Robinson, 2008

Stephan Schmidheiny, *Changing Course: Global Business Perspective on Development and the Environment*. MIT, 1992

E.F. Schumacher, *Small is Beautiful: A Study of Economics as if People Mattered*. Blond and Briggs, 1973

Lynn Sloman, *Car Sick: Solutions for Our Car-Addicted Culture*. Green Books, 2006

Alex Steffen (ed.), *World Changing: A User's Guide for the 21st Century*. Harry N. Abrams, 2006

Nicholas Stern, *The Economics of Climate Change: The Stern Review*. Cambridge University Press, 2007

Nicholas Stern, *A Blueprint for a Safer Planet: How to Manage Climate Change and Create a New Era of Progress and Prosperity*. Random House, 2009

Oliver Tickell, *Kyoto2*. Zed Books, 2008

Gabrielle Walker and David King, *The Hot Topic: How to Tackle Global Warming and Still Keep the Lights on*. Bloomsbury, 2008

Worldwatch Institute, *Vital Signs: The Trends that are Shaping our Future*. Norton, published annually

Joanna Yarrow, *How to Reduce Your Carbon Footprint*. Duncan Baird Publishers/Chronicle, 2008

USEFUL WEBSITES

Association for the Study of Peak Oil and Gas (www.peakoil.net)

Audubon Magazine (www.audubonmagazine.org)

C40 Cities (www.c40cities.org) – the world's largest cities coming together to fight climate change

Carbon Footprint (www.carbonfootprint.com) – features a carbon footprint calculator

Celsias (www.celsias.com) – forum for environmental news and debate

Climate Debate Daily (www.climatedebatedaily.com)

Clinton Climate Initiative (www.clintonfoundation.org/what-we-do/clinton-climate-initiative)

Compassion in World Farming (www.ciwf.org)

Conservation International (www.conservation.org)

Earth Day Network (www.earthday.net)

Earthwatch Institute (www.earthwatch.org)

The Ecological Farming Association (www.eco-farm.org)

The Ecologist online magazine (www.theecologist.org)

Freecycle (www.freecycle.org) – nonprofit network for giving and receiving unwanted property for nothing

Friends of the Earth (www.foe.org)

Green Building Council (www.usgbc.org or www.ukgbc.org)

Green Car Congress (www.greencarcongress.com)

Grist (www.grist.org) – online magazine of environmental news founded by the Earth Day network

Happy Planet Index (www.happyplanetindex.org) – measures the ecological efficiency with which human well-being is delivered

International Vegetarian Union (www.ivu.org)

LETSystems (www.gmlets.u-net.com) – advice on developing a Local Exchange Trading System

Live Earth (www.liveearth.org) – following on from the Live Earth concerts in July 2007, a great online tool providing information on climate change and reducing carbon emissions

Man in Seat 61 (www.seat61.com) – plan flight-free journeys

Marine Stewardship Council (www.msc.org)

National Geographic (www.nationalgeographic.com)

Off-Grid (www.off-grid.net) – advice on reducing your home's dependence on mainstream energy and water supply services

The Oil Drum (www.theoildrum.com) – digest of news and debate on energy-related issues

Oxfam International – Make Trade Fair (www.oxfam.org/en/campaigns/trade)

Renewable Energy World (www.renewableenergyworld.com)

Rocky Mountain Institute (www.rmi.org) – nonprofit organization that "fosters the efficient and restorative use of resources"

Scientific American (www.sciam.com) – news, articles and information on all aspects of science, including environmental issues

The Soil Association (www.soilassociation.org) – promotes sustainable, organic farming

Urban Gardening Help (www.urbangardeninghelp.com)

Treehugger (www.treehugger.com) – in-depth blog-style website about green lifestyles

The Vegetarian Society (www.vegsoc.org) – an educational charity promoting understanding and respect for vegetarian lifestyles

World Resources Institute (www.wri.org) – an environmental think-tank that goes beyond research to find practical ways to protect the Earth and improve people's lives

World Wildlife Fund (www.wwf.org)

World Wind Energy Association (www.wwindea.org)

Worldwatch Institute (www.worldwatch.org) – a leading source of information on the interactions among key environmental, social and economic trends

INDEX